To my dear Dad,
Keep the blue side up!
Love, Sean 2004

FLIGHT PATH

FLIGHT PATH

HOW **WESTJET** IS FLYING

HIGH IN CANADA'S MOST

TURBULENT INDUSTRY

PAUL GRESCOE

John Wiley & Sons Canada Ltd.

National Library of Canada Cataloguing in Publication

Grescoe, Paul, 1939-
    Flight path : how WestJet is flying high in Canada's most
    turbulent industry / Paul Grescoe.

Includes index.
ISBN 0-470-83436-6

    1. WestJet Airlines. 2. WestJet Airlines—History. I. Title.
HE9815.W48G74 2004   387.7'06'571   C2004-901480-3

Production Credits:

Cover design: Ian Koo, Adrian So R.G.D.
Cover photo of plane: Creative Intelligence Agency
Interior text design: Adrian So R.G.D.
Printer: Tri-Graphic Printing Ltd.

Printed in Canada
10 9 8 7 6 5 4 3 2 1

DEDICATION

To Audrey,
my skilled and supportive co-pilot
on the flight of our lives

And to Taras, Justin, and Lara,
who are winging
their own unique ways
through life

CONTENTS

ACKNOWLEDGEMENTS

Like WestJet itself, this book was a co-operative enterprise and I am grateful to all those who participated in its making. Prominent among them are the four co-founders—Clive Beddoe, Don Bell, Mark Hill, and Tim Morgan—and their colleagues who agreed to give me access to anyone I wanted, and all the people of the airline across the country who were so forthcoming in interviews. My first and continuing contacts at the company deserve special thanks: Bill Lamberton, Siobhan Vinish, and Sarah Deveau. A couple of key executives were of great assistance: Sandy Campbell, who offered a lot of informed background on people, and Stu McLean, who was always handy with an apt airline analogy. And, outside the company, I relied on the close and keen observation of the Canadian airline industry by several journalists—including Keith McArthur, Paul Vieira, and Peter Verburg—to put WestJet in context.

There were many others who contributed. Allan Bergman of ShowMakers Productions, Deepak Sahasrabudhe, and Doug Fleetham generously made the first attempts to rescue the flawed tape of a vital interview and Larry Baker and Lynn Asselin of Vancouver's Headroom Studios finally restored it with their skills. Ann Douglas transcribed all the tapes with diligence and intelligence. Of course, the book wouldn't have happened without my knowledgeable (and very nice) agent, Carolyn Swayze, and the publisher who recognized its potential, Karen Milner. Cheryl Cohen astonished me with her copy-editing and fact-checking abilities. And, as always, Audrey Grescoe was literally there for me in every way during the on-site research phase and took over the controls of another important project while I was completing this one.

PROLOGUE
Graduation Day

"From the makers of Sleepless in Sault Ste. Marie, Days of Thunder Bay, and Good Morning Victoria comes the story of a vision for a different kind of airline. One that would fly only jets, that wouldn't rip people off, and that would always remember to smile ... This is the story of WestJet."
> —Narrator of a corporate video in the overblown style of a movie trailer

This is a book about a Canadian company that flies—literally—in the face of conventional business wisdom, against the prevailing winds of its industry. A company that says the customer doesn't come first and isn't always right. And yet one that respects its customers and insists on calling them "guests" while officially labelling its accounting department "BeanLand" and its executive team "Big Shots." An enterprise that, from the beginning, decided it didn't have to reinvent the joystick to soar in the marketplace, that it could borrow brazenly from fellow high flyers—and brag about it. That focused not on the canon of market share but on the more creative principle of market stimulation to create a whole new base of customers. That still prefers to hire people not for

I

their relevant experience but for their passion and positive attitudes and then trains them in the skills they need. That gives all its employees a fair share of the profits and matches their purchases of the company's stock, dollar for dollar, to turn some of them into millionaires. And that, after making them co-owners, gives them the power to micromanage their own company, to make immediate decisions that keep costs low and productivity high (with pilots helping clean aircraft cabins and customer-service agents doubling as flight attendants). That encourages a non-union, self-run employee organization to act as a corrective to any wrong-headed management practices and welcomes one of its representatives as a full member of the corporate board of directors.

A company with low costs and fares, good turnaround times, and bad inflight jokes. And that, at this writing, has never had less than a profitable quarter in eight years and now jets sky-high and safely above the wreckage of the Canadian—and global—airline industry.

The story of WestJet Airlines has taken on near-mythic overtones: how this tough little bird controls its own destiny in a highly regulated industry; how it once "hijacked" an airplane from the United States; how, during a battle with Ottawa, it shut itself down voluntarily and lost millions of dollars. Of course, even legendary companies can have Achilles heels and certainly WestJet has suffered internal stresses and strains, made big mistakes, and encountered severe turbulence. The challenges have involved the antics of a relatively few executives and supervisors who didn't fly the idealistic flight path, as well as the failures of basic communication that encourage unions' attempts to seduce employees into their embracing bosom.

In spite of the bumps, WestJet has got enough of it right to loom as a fascinating role model for managers in other industries, employees in other fields. An airline analyst recently called the carrier "the only star in a universe filled with black holes." A professor of economics described it as "one of the most successful airline start-ups in history." And, most tellingly, a senior business journalist—in profiling "the fastest-growing airline ever launched in Canada and one of the most profitable in North America"—credited much of its success to the corporate culture that allows its people to manage the company from the bottom up.

There are obvious lessons to be learned from a flourishing business venture that in 2002 was selected as one of Canada's Top 100 Employers (in a book of the same name) and in 2003 the nation's second-most-respected corporation, after the Royal Bank of Canada, and most respected for customer service and high-quality service/product (according to an Ipsos-Reid poll of CEOs). And from a quartet of company founders who in 2001 were named the top Canadian entrepreneurs and went on to Monte Carlo to win the teamwork award in Ernst & Young's first World Entrepreneur of the Year competition. WestJet Airlines is one of the few recent unassailable success stories in the Canadian business community that is worth telling, warts and all. In an industry where twenty-eight domestic airlines have disappeared in the past twenty-five years, how has WestJet stayed aloft?

The fuselages of all WestJet airplanes show a Canadian flag with "Proudly" above it. The forward bulkheads bear the slogan "Way To Go!" The latter phrase carries a couple of meanings: it suggests that this is the airline to fly and it's also a congratulatory exclamation that conveys the spirit of the company. But the three little words can have a larger implication. They're another way of saying that WestJet is a worthy role model to emulate—or, in the jargon of its industry, a proven flight path to follow.

3

GRADUATION DAY

What were all these folks—of various ages but following the same seriously casual dress code—doing in the conference room of a suburban Calgary hotel, saying highly uncomplimentary if very funny things about Air Canada? If I hadn't known, I couldn't have guessed why these fifty-five men and women had gathered together on a broiling August afternoon at the Coast Plaza Hotel. But even a casual passerby would realize that this was no conventional business conference. WestJet Airlines was mounting another Culture Presentation, a regularly scheduled manifestation of the corporate vision, values, and mission for its employees—whom the management insists on calling "people" rather than the distancing "personnel," "staff," or "employees." Many of the young women in jeans and sleeveless shirts were hardly out of their teens, while a handful of men in shorts and T-shirts were well into middle age. On the job, some would sport unpretentious airline uniforms (no pilots' caps), but most dressed down in the relaxed clothing they were wearing here (not a suit or a tie to be seen). Their accents ran from light Alberta twangs through flat Ontario tones to rich Newfoundland lilts. Almost all the employees were newcomers, save for the occasional veteran who had somehow managed to miss an earlier presentation. They included reservations agents and information technologists, ramp workers and pilots, experienced aircraft dispatchers behind the scenes and green customer-service staff on the frontlines in airports across Canada. There was at least one second-generation WestJetter: Kim Jordan, the boisterous young daughter of Shelley Jordan in the airport standards and procedures group, would soon be one of about a dozen women amid the 136 ground crew humping baggage on and off 737s at the airport in Calgary, the airline's home base.

4

A current of anticipation buzzed through the room. These graduates of in-house training courses gathered around tables they'd selected according to various self-descriptive labels perched on top: "Passionate," "Innovative," "Friendly." Other signs reminded them that "CARE is Creating A Remarkable Experience." The customary earnest stuff of corporate pep rallies—or was it more than that?

The lights dimmed and a video flashed on a large screen. It looked like a slightly wonky movie trailer—"The following airline has been approved for all audiences"—with mock-heroic music and voice-over that set the serio-comic tenor of the six-minute documentary. The opening scene has close-ups of hands gesturing around a boardroom table littered with Pepsis and pizza while people say "So ... we'll start with three aircraft," "Give me one good reason why this isn't going to work," "It's not that difficult a business, to get people from A to B." A narrator declaims: "It is a tale of kindness, goodwill—and a good old-fashioned sense of humour." Cut to a female customer-service agent on a microphone: "At this time, let's commence our pre-board, and guests travelling with small children, or guests thinking they are children, please ..."

The portentous narrator invites the audience to "come back with us to the early days when excitement and high spirits took an unexpected body blow—a voluntary grounding just six months into operation." There, standing before the engine of a 737, is the president and CEO of WestJet, Clive Beddoe: "Let's be absolutely clear. It was WestJet that decided to put this airline on the ground. We decided to take control of the situation to make sure the public at large had confidence in our airline. But it was a defining moment in the history of our company."

Later, there's a scene at the Calgary International Airport on September 11, 2001, when WestJet people leaped in, unbidden, to help care for hundreds of foreign passengers from four international airlines diverted to the city in the wake of the terrorist attacks in New York and Washington. "This is a story of courage ... a story of never looking back ... a story of believing in yourself—no matter what ... Starring you ... Now playing across Canada." The sound-track swells to a corny, rousing climax.

As the lights came on in the conference room, a kindly-looking, sandy-haired man in his late forties strolled to the middle and said, "Hello. I'm Don Bell. I was one of the guys sitting around the table eating pizza as we were planning an airline." Before becoming one of the four founders, Bell had been the owner of a successful computer software company and an experienced commercial pilot who occasionally flies for WestJet to keep his hand in. Only that morning he'd flown from Calgary to Abbotsford, British Columbia, to fill in for an ailing pilot, and when he made a public-service an-nouncement the passengers gave him an ovation. His day job is co-chief operating officer as well as senior vice-presi-dent, customer service, a role that put him in charge of perpetuating the WestJet culture as the company grew from 220 people to more than 4,300.

Don Bell is not someone who smiles a lot, nor orates with the histrionics of a professional motivational speaker. At various points in his presentation, though, he does utter pithy sayings in rhyme: "Stop that stinkin' thinkin'!" and "Get a checkup from the neck up!" But in his plain-spoken passion, he wafts sincerity. (A male flight attendant, hearing I might attend a Culture Presentation, remarked without irony, "It's almost a religious experience.") Bell is the self-appointed torchbearer of the corporate cul-ture, what came to be called the WestJet Spirit or even WestJettitude.

Relaxed this afternoon in a short-sleeved sport shirt, hands in pockets, he was asking the new employees why they'd wanted to join the company: "How many people knew somebody at WestJet? What did they say about WestJet?"

Fifteen hands shot up. "Company that cared." "Fun place." "Good future." "Profitable."

"You must be from Canadian Regional," Bell joked with the last speaker, harking back to the defunct airline from which many of WestJet's earliest employees had escaped. Casually, he began recounting a potted history of his company, The Little Airline That Could: "We didn't know anything ... Where do you get airplanes? How do you sell tickets? We didn't have preconceived notions of how it should work ... We started piggybacking on some of the ideas of Southwest. In an industry where you have the worst disaster, you also have the best success: Southwest Airlines."

The most successful carrier on the continent, Southwest was one of two pioneers of the low-cost, low-fare, people-first model in the United States (the other was Pacific Southwest Airlines). Over three decades, its common stock became the best investment on the American market, with Wal-Mart trailing a distant second. In recent years, financial analysts have valued it higher than all other major US airlines combined. Its secret, which WestJet borrowed without blinking, is to nurture a culture that emphasizes positive relationships among its employees—in an environment that embraces both mutual respect and a little wacky humour to leaven the hard work its people perform.

Don Bell mentioned David Neeleman, who a decade ago sold his small Salt Lake City airline to Southwest after developing Morris Air into another lucrative low-fare carrier and introducing the industry's first fully ticketless system and its simplest reservation system. WestJet adopted both innovations

7

after Neeleman advised and invested in the new Canadian carrier. Now in his early forties, worth well over $200 million, he runs New York–based JetBlue, an airline that Fortune Small Business magazine says is so choosy about picking customer-sensitive staff that "getting hired is harder than being accepted into the Ivy League" (in 2002, 130,000 people jockeyed for two thousand jobs).

That's the kind of company Don Bell aspired to help build back in 1995, when he, and co-founders Tim Morgan, Mark Hill, and especially Clive Beddoe—the financial and philosophical driving force—dreamed up their airline. As Bell recounted, WestJet launched on February 29, 1996, with great success, only to be harassed by Transport Canada bureaucrats who did a sudden flip-flop over regulations for aircraft maintenance manuals. In Bell's mind, the reasons behind the campaign against this upstart airline were highly suspect. So, rather than being pecked to death, Beddoe decided late one night to shut the operation down, ground the airplanes voluntarily, and work to meet all the new demands in a concerted campaign on their own terms. "The first decision we made was not to lay anybody off." The whole staff pitched in to satisfy the regulations—Bell's wife "was in charge of the refund queue; we called her the Refund Queen." When WestJet opened its phone lines to resume service seventeen days later,"we expected about five thousand calls that day. We got fifty thousand," he said a little hyperbolically. "And we've never looked back."

With WestJet's stock-option, share-purchase and profit-sharing plans, many of the original pilots have become millionaires and the initial classes of flight attendants who bought shares have earned hundreds of thousands of dollars as the stock split twice. "We've created a bunch of filthy capitalists," Bell joked. Yet here in the hotel conference room, the unspoken question hung in the air: "How well can we newcomers do all these years later?" He assured them: "You're getting in on the ground floor. The gravy train isn't the past six or seven years. It's actually starting ... The airlines that are doing well are just kicking butt."

Those would be the low-cost carriers like WestJet, including Southwest and JetBlue in the United States and Ryanair and easyJet in Europe. But, as Bell well knew, 2003 was far from over and most of the airline industry around the world was already suffering an annus horribilis. The fallout of the war in Iraq, the outbreak of Severe Acute Respiratory Syndrome (SARS), and even the lingering legacy of 9/11 had battered business and leisure travel. The reality of the downturn surfaced in the rotten news the media had been reporting during July. The Dutch carrier KLM announced a full-year loss of almost half a billion dollars (US). Singapore Airlines, forced to slash fares by as much as half to lure SARS-wary passengers, was forcing its staff to take a week's unpaid leave every two months. At one point, Cathay Pacific Airways of Hong Kong cancelled more than 40 per cent of its flights. Qantas of Australia let go a thousand employees. In the United States, United Airlines struggled through Chapter 11 bankruptcy protection after a first-quarter loss of $1.3 billion (US), while Continental Air Lines deferred a $2.5-billion order for thirty-six Boeing 737s for at least five years. And here at home, Air Canada was desperately trying to restructure itself and emerge from bankruptcy protection by seeking $700 million (Cdn) worth of investment to save the world's eleventh-largest airline.

Against this dismal backdrop, nervous WestJet shareholders had been holding their collective breath and business analysts were primed to pounce on the airline's second-quarter results. The betting was that WestJet would

9

finally post its first-ever loss. "The overwhelming mood on the street," said Cameron Doerksen of Dlouhy Merchant Group Inc. in Montreal, "was fairly pessimistic." But just two weeks before this Culture Presentation, WestJet reported its twenty-sixth consecutive profitable quarter with a 20-per-cent increase in profits to a second-quarter record of $14.7 million. The analysts were obliged to admit they were pleasantly surprised: The news propelled the airline's share price to its biggest single-day gain as the stock climbed by nearly 20 per cent to close at $22.09 on the Toronto Stock Exchange. (By September, the share price would have risen by nearly 50 per cent and Doerksen would be saying "WestJet has operational metrics that compare very favourably to its closest US peers, Southwest and JetBlue. It is on this basis that we believe [it] deserves a multiple that is more in line with these stocks.")

Bell pointed out that if WestJet grew 30 per cent a year, its stock could double every two and a half to three years. "And that's realistic because we've been growing at 50 per cent a year." He knew that for this audience, the prospect of being able to sock away up to 20 per cent of their pay into WestJet stock—and have the company match the contribution—was vital information. The airline is not known for paying high basic salaries: customer-service agents start at $10.45 an hour and first officers at only $45,000 a year.

Someone in the room wondered: "If we're growing at the rate of three people a day, how do we ensure we maintain WestJet as a winner?"

The WestJet VP told of having to fly recently on Horizon Air, Alaska Airlines' regional subsidiary. When the flight was delayed, a customer-service agent told him, "What do you expect of Horizon?" To the new employee, Bell said, "It's up to us. Where does the attitude come from? ... It's defined by the actions of the management. I think this will always be a small airline. It will just have a lot of people working for it. You know, all we have to do is not screw it up."

The potential for screw-ups in his industry is immense. He showed a slide that compared airlines' operating expenses: Air Canada's were $333.82 per passenger, Southwest's $121.67 (Cdn), and WestJet's $101.82. Another slide listed twenty-nine Canadian carriers that no longer existed. One of them was Roots Air, which Skyservice Airlines ran for six weeks in 2001, losing an estimated $7.5 million before Roots shut down.

Throughout his presentation, Bell kept circling back to the core values that he and his partners had adopted early on and that have to be sustained to keep the airline profitable. Hire the right people—glass-half-full people—and help them to think like owners. Make sure the corporate language reflects the culture: Sales Super Agents versus reservation agents, Team Leaders versus supervisors, Promises versus policies ("in customer service it's easy to hide behind a policy"). And allow WestJet people to express themselves openly, authentically: "When you come to work, you don't have to put on a mask—you just be yourself."

After a break for coffee and cookies, the new employees had a chance to be themselves. An enthusiastic new facilitator, Joanne Leskow, asked them to concoct impromptu skits, songs, and poems to reflect the corporate values they'd been absorbing. (Such was the zeal for this interactive role-playing that some people at my table urged me to join them in the creative process. I demurred and was grateful I'd done so when names like Michael Jackson, Village People, and some rap singer were tossed around as inspiration. One of the young women sat silently with her arms crossed during the fifteen minutes of conception; I suspected her future at the fun airline was not assured.) Predictably, most of the skits revolved around how uncaring Air Canada is compared to WestJet. As if to underline that sentiment, Don Bell then announced—to cheers and applause—that the token rate the company charges employees to fly anywhere in Canada had just been lowered to $2.50 from $11 a trip.

Now he was introducing a soft-sell corporate video that telegraphed even more of the culture. It cut between groups of happy WestJetters and iconic images of Canada from the air—the Peace Tower, mountain peaks, herds of wild horses. If the message was equally uplifting, and perhaps platitudinous, this audience was ripe and ready to fly with the video's message: "We are positive and passionate about everything we do. We believe service comes from the heart, not from a manual. We are friendly and caring and treat everyone with respect. We embrace change and innovation. We empower people to find solutions. We are honest and open and keep our commitments. We take our jobs seriously, but not ourselves." That last line featured a scene of the airline's president flashing his 150-watt smile at a couple of elderly female passengers—followed up by a shot of a male flight attendant in drag.

"We are Team WestJet," the video concluded. And waiting in the wings just outside the conference room, about to make his promised appearance, was a man most of the people in the room hadn't yet met, the man who personifies WestJet in the public's mind, the head coach of a team that just keeps winning: Clive Beddoe.

The businessman who once (briefly) believed that launching a low-cost airline in Canada was a lousy idea.

Not to mention the guy who had appeared in the airline's latest glowing annual report with sleeves rolled up, unloading a dishwasher, in a photo captioned: "He makes one heck of a burger too!"

And the corporate leader who in 2004 would be ranked in a national poll of his peers in other major companies as one of the top dozen most-respected CEOs in Canada.

1

WIN YOUR WINGS FROM THE GROUND UP
The guys behind WestJet

Wayne says the WestJet flight attendants had some great one-liners during his recent trip to Abbotsford, but this one really made him laugh:

"At this time, we'd like to invite all smokers onboard to join us upstairs in the smoking lounge for our inflight movie, *Gone With the Wind*."

—People column, *The Edmonton Journal*

Frankly, my dears, unlike Rhett Butler he *did* give a damn. Clive Beddoe gave a damn about flying, he gave a damn about fleeing Britain's bureaucracy, and two decades later he gave a damn about finding a way around the ridiculously high cost of getting from Calgary to Vancouver on a commercial flight.

Beddoe's first thrilling experiences of flight came while winging at low altitudes in a flimsy glider catapulted across the green playing fields of Epsom College in England. Those teen-years aerial adventures were enough to ignite a lifelong passion that would propel him to earn a licence in the early 1970s to fly airplanes and helicopters—but always as a private pilot who never seriously considered aviation as a career.

Is it any coincidence that in the year Clive was born—1946—British transatlantic airline service began to North America with Lockheed Constellations crossing the pond in nineteen hours, forty-five minutes? Or that Scandinavian Airlines, Alitalia, and Air India were all born in the same year? Or even that the English director David Lean released his prophetically titled film *Great Expectations*? Were these all auguries for the boy born in that fateful year?

Clive grew up in Leatherhead, a small, ancient market town in Surrey south of London. With the River Mole meandering through and greenbelt countryside surrounding it, this is an officially designated Area of Outstanding National Beauty. But as he matured, Clive couldn't wait to flee his British birthplace for greener pastures overseas. Roughly handsome, with a broad nose and the hooded eyes of a pilot who's seen too much sun, he was the second eldest of three sons and a daughter. His father, Ken Beddoe, had settled into the security of the civil service after serving in the war as a navigator with the Royal Air Force. His mother, Kay, was a more creative thinker and more ambitious; together, they bought a large Victorian house and converted it into a rental property. The income allowed Clive to become a day pupil to a nearby boys' school at Epsom, a town twice the size of Leatherhead and famous for its healing salts and derby. Epsom College, founded in 1855 under the patronage of Queen Victoria, ranked second in the A-level league of England's public schools. "And that set me up," he reminisces in his soft, lingering English accent. "I spent a lot of my influential years amongst fairly affluent people—much more affluent than we were, but also they came from perhaps more entrepreneurial backgrounds."

The college had a cadet corps. Swayed by his father's background and his own fascination with airplanes, Clive

16

chose the air-force section. The cadets could fly the school glider, and his joy aboard the engineless craft led him to a summer glider camp and more flying after school. His father's interest in property management may have influenced his decision following public school to article for a firm of chartered surveyors—land and building assessors—while taking evening courses at the College of Estate Management. Part of London University, the college taught the art and science of property evaluation four hours a night in a four-year program. On graduating, Clive Beddoe became an associate member of the Royal Institution of Chartered Surveyors, the home of a worldwide network of property professionals.

That was the end of his formal education. Book learning is all well and good—it can suggest a flight path and how to follow it—but Beddoe, like the other three founders of WestJet Airlines, never pursued an advanced degree, much less an MBA. They all earned their stripes in the real world, working at a series of jobs, a progression of positions, and eventually, in three cases, building their own entrepreneurial enterprises that predated WestJet. And although the four shared a love of flying, only one had made aviation his profession. Over many decades they had, however, assiduously done the down-to-earth spadework in their varying pursuits that would let them take wing together with an astonishingly successful, against-all-odds airline. It was their combination of talents and—this is the secret—their worldly experience that came neatly together like a finely crafted key slipping into the right lock.

Back in the late 1960s, Clive Beddoe's position in the chartered surveyors' firm that employed him could not have been less entrepreneurial. While he might become a junior partner within a couple of years, he could easily wait a quarter of a century before making senior partner. "I became very frustrated by the system in England, which is basically that you succeeded by following in dead men's shoes. So your career was managed for you with absolute certainty from that point on. What an awful existence. Certainly there would be financial success, but the challenge was taken away."

Beddoe had always been intrigued by those who created and built their own businesses, Britain's environment was not particularly nurturing to such self-starters. It didn't take long before an entrepreneur reached the 90-per-cent tax bracket. "There was no incentive, nobody cared, you couldn't get people to do things because there was no reward in it. And the union movement was so unbelievably destructive. It was an impulse to leave as much as an impulse to come to Canada."

He'd entertained four possible destinations. South Africa, despite its wealth, was a statistically dangerous place to live. Australia/New Zealand were warm and welcoming, but a long way away. The United States, which might be difficult to get into, was also a bit intimidating because of its storied cutthroat commercial environment.

"And then there was Canada, and Canada really appealed to me because it was a country with big, wide-open spaces and not a huge population base. Obviously a country of the future, something between the US and Britain—with what I thought would be a business environment with enough of the US influence to be commercialized, but not overly influenced by the British attitude."

Getting there would be a problem because he'd need the offer of a job somewhere, anywhere, in the country to amass enough points to qualify under the immigration system. Then one of three friends living with him in London was hired by a recruiter from Knowlton Realty in Calgary who visited England each year to find chartered surveyors because they were in short supply in North America. When he was about to finish his final exams at the estate-management college, Beddoe wrote to his friend in Canada to help find him a job there. "I just wanted out," he says now.

In late November 1970, the friend called him. "If you can be in Canada in a month, I've got a job for you."

"Done."

"Don't you want to know what the job is?"

"No."

After the would-be emigrant tendered his resignation at work, the senior-most partner of the firm, a war hero, summoned him into his presence and said: "Beddoe, you're a traitor."

He sold all his belongings and arrived in to Calgary with $999, which went promptly into a bank account. One of his first financial transactions was an inauspicious introduction to his adopted country. As he was buying a cassette recorder at Eaton's, the sales clerk refused a cheque from this new immigrant, yet signed him up for a department-store credit card. "If I've got a credit card now," Beddoe said, "can't I write a cheque?" He could. The cheque bounced, however, because when his nest egg was wired from England, the bank had put a decimal point in the wrong place and credited his account with only $99.99. Welcome to Canada, Clive.

His friend had already left Knowlton to work in property management for Marathon Realty, an arm of Canadian Pacific Railway. He asked Clive to help out. "I lasted there three months. It was awful, so absolutely bureaucratic." Shopping around for the most interesting developer in town, Beddoe heard about Alan Graham of the Cascade Group of Companies. Knowing the man was a Scot who might understand the weight of a chartered-surveyor's diploma, he phoned this very circumspect operator of a privately held conglomerate: "I think you should hire me."

"Why?"

"Because I think I can make you some money." Welcome to Clive, Canada.

Cascade owned real estate, wax museums, and nursing homes in western Canada. In Calgary, Graham ran a construction company and developed apartment buildings. "He had lots of gross revenue, but not much on the bottom line," Beddoe remembers. Over the next seven years, as the new development manager, Beddoe took Cascade into the more reliable sector of office buildings, among them the Norcen Tower, the Exchange Tower, and the Sun Oil building in downtown Calgary, acquiring the land, negotiating permits, overseeing construction. Beddoe was eventually able to negotiate a financial participation in these developments and his piece of the pie grew big enough for him to branch out on his own. Observing that Cascade was beginning to operate on the edge of his "moral comfort zone," Beddoe had demanded his money but had to settle for two-thirds of what he was owed, about $220,000 net. "Ultimately Alan Graham left Canada."

In the early 1980s, Beddoe started his own commercial-development company with $100,000. He named it Hanover Management for London's Hanover Square, where he'd once worked—"it sounded royal and as if it had some money behind it." His partners in the venture were oilman Harley Hotchkiss and well-placed local lawyer and Liberal Jim Palmer. Hotchkiss owned Sabre Petroleums with the colourful German Baron Carlo von Maffei (who had a 23,000-square-foot house outside Calgary protected by armed patrols) and later became a partner in the National Hockey League's Calgary Flames. Hanover specialized in buying older office properties downtown, including the historic Lougheed Building, as a way of holding and controlling land for such developments as the 220,000-square-foot Hanover Building, eventually leased by Petro-Canada.

After the federal government's National Energy Program (NEP) of 1980 led oil companies to downsize or move out of town, Calgary's real-estate market cratered. With continuing cashflow from his properties, but deep in debt to his bank, Beddoe looked around for other opportunities. He borrowed further to buy Career College, a private vocational school in Calgary and Edmonton that was thriving during recessionary times. There were parallels with the kind of company he and his partners would later create in the airline industry. His female president developed what he calls "a nurturing environment" and every student the school graduated beyond a certain break-even point produced almost pure profit—"just like filling another seat on an airplane." Unlike its competitors at the time, his school's strategy was job placement rather than education for its own sake. "To place people in jobs, you had to educate them the way the industry wanted. So you had to understand what they needed in order to modify your syllabus and teach the skills the people wanted to get a job." The challenge, he notes now, is not so different from analyzing and satisfying airline passengers' needs. Expanding to Toronto and Ottawa, he owned Career College for fourteen years, most of them profitable until the government cut back on student loans. (And it was through the school that he met his wife Ruth, who was in charge of finding jobs for the graduates.)

During the 1980s, Beddoe was developing the Hanover Group of Companies, which came to include a 50-per-cent investment in Canadian Paper Recycling of Calgary, a money-losing company that he turned around before selling out. In 1994 he diversified again, acquiring Western Concord Manufacturing, a plastics company with plants in Edmonton and Vancouver. Keeping tabs on the business demanded almost weekly flights to the west coast, usually on Air Canada, at $700 a round trip. There had to be a cheaper way to fly the triangle of cities. He consulted Tim Morgan, a friend who ran an air charter and flying school and to whom Beddoe had leased his first airplane, a small Cessna 172. With Morgan's help, he bought the larger, more powerful Cessna 421, an eight-seater that Beddoe began flying to Calgary, Edmonton, and Vancouver, cutting his commuting costs by about two-thirds.

THE FLYING FARMBOY: UNIONS VS. AIRLINES

In the early 1990s Tim Morgan was a pilot for Canadian Regional, a subsidiary of Canadian Airlines International, who operated Morgan Air Services on the side. Then approaching his forties, he had an open country-boy's face and a bull's shoulders, reflecting his childhood on a cattle and grain farm near Strathmore, Alberta (which he and his family still own today). The town, just east of Calgary in the county of Wheatland, was a creation of the Canadian Pacific Railway, a landing pad for nineteenth-century immigrants coming to learn how to farm the rich prairie soil. As a teenager, Tim was less enamoured of agriculture than aviation. He got the flying bug—"I think it was from driving tractors around in circles in a field instead of going in a straight line. Lots of farm boys fly, they're a bit mechanically inclined, kind of like the open skies." His first job off the farm was polishing airplanes at an aircraft sales and marketing company in Calgary.

After high school, he studied aviation at Calgary's Mount Royal College and, with two other pilots (who would later work at WestJet), bought a Citabria Decathlon, a two-seater aerobatic plane, built a hangar on his farm, and built up his flying hours. Being a pilot was an escape from the land, in more ways than one. "It's not that I disliked the farm, it's just a very difficult lifestyle, a lot of hard, hard work." After he earned his flying instructor's rating at a Calgary flying school, his first job as a nineteen-year-old was teaching rural people how to fly on small local airstrips. His next one was flying a summer and a half for a game outfitter along the legendary Canol Road in Yukon in a single-engined Piper Super Cup with tundra tires. The idea was to spot big mountain sheep for hunting expeditions: "Then the guide would take the hunters on a big, long tour all over the country and finally get to the place where in fact we knew the sheep were." He punctuates the story with his easy laugh.

Morgan soon graduated to larger Cessna 401s for a Calgary charter airline that served the oil patch in Alberta and a year later spent a summer seeding thunderclouds on hail-suppression flights in a twin-engined Cessna 320. It sounds dangerous, he admits, but it's not—"I don't have that many horror stories to tell you about aviation." By then, ready to settle down, he became a corporate pilot for a company selling road-construction equipment. "It was one of the better corporate jobs around then," he says. "The owner respected you; you weren't just a taxi driver. And I knew the intricacies of the company." In the early 1980s, the energy crisis and the launch of the NEP shook the Alberta economy. "I was one of the last corporate pilots in Calgary to lose their jobs. I remember when I first started, I couldn't find a place to park the airplane on the ramp. And then when I finished, I used to shotgun down that ramp and not hit anything."

Meanwhile, he'd saved enough money to buy his own four-seater Cessna and rent it out. One of his lessees was the operator of the school where he'd earned his instructor's rating. When the man decided to leave town, Morgan inherited his clientele and set up his own flying school, Morgan Air Services. He hired an instructor from Eastern Canada who helped him build the business, which evolved into a charter operation that today has fifteen aircraft, from light twin-engines to the reliable workhorse King Air. It has been an efficiently run company; "Tim likes the standard operating procedure and everything by the book and written down," says a friend and fellow businessman.

After a couple more corporate-pilot positions, one of them flying a Gulfstream Commander turboprop, Morgan had an unexpected call from Time Air, which had begun as a small commuter airline in Lethbridge, Alberta, and expanded to cover most of Western Canada. The caller wondered if Morgan would like to take a Fokker F-28 course—the following day—and then come fly the sixty-five-seat jets for the company. "The next thing I knew, I was a co-pilot on this big airliner and I looked out the window and where am I landing? In northern Saskatchewan on a gravel S-curve with a fishing shack—doing the same thing I always did."

Time Air was, in many ways, a model company. "People wanted to do a job, make things happen, they wanted the company to be successful—and they just wouldn't come to work and go home ... There was profit-sharing. It was very fun, very friendly." For the next two and a half years, he worked out of Saskatoon with the former employees of NorcanAir, which Time had taken over. "That was a very tight-knit group, very work-oriented, very—I hate to say this—very Alberta agricultural-type people. I fit right in there."

But Time, and the times, were a-changin'. With the deregulation and consolidation of the domestic airline industry in the late 1980s, the privatized Pacific Western Airlines bought Time Air and Canadian Pacific Air Lines (CP Air). Time became Canadian Regional, a feeder airline to Canadian Airlines International—the renamed CP Air. In the early 1990s, Tim Morgan wound up back in Calgary with Canadian Regional as a captain on de Havilland Dash 8 turboprops. "And that's when the airline business started to come apart," he recalls.

Morgan joined the Canadian Air Line Pilots' Association (CALPA) and became vice-chair of the council representing the entire regional arm of the airline. "We thought we had a good relationship with the management within Time Air. But we got sucked into the old mother CALPA, which was Air Canada and Canadian at that time. And we were such a small fish in a big pond, we got told what to do. To the point we went on strike."

Then he met what he considers the true face of unionization in his industry: "We were to sign a contract regarding the merger between one of the companies down east and Canadian Regional. And we went down there saying 'our pilots don't want to sign this, we want to re-look at it.' We walked into the hotel room in Toronto and sat across the table from the CALPA lawyers and this guy—great big cigar, greasy beard, from down east—just what your typical union guy would look like. And he says, 'You must sign the contract. It really doesn't matter whether you want to sign it or not, if you don't sign it, we'll sign it for you.' So I said, 'Oh, okay.' Picked up my briefcase, walked out of the hotel room, got back in the elevator, out to the airport, on to the airplane, came home, and that's the last thing I ever had to do with CALPA. It was very much a learning experience for me, because the people asked me to represent them and I couldn't represent them. This was nothing but a sham. Why

25

was I involved in the union? Because if someone was gonna run my life, I thought it might as well be me."

While Morgan survived the layoffs caused by deregulation, he was bounced around from the senior officer's position on the F-28 to mid-level captain on the Dash 8 and back again. But his schedule afforded him enough free time to keep running Morgan Air Services—where one day in 1994, Clive Beddoe showed up with Mark Hill, one of his employees at the Hanover Group.

THE STRATEGIST: THE LURE OF LOW COST

Mark Hill, a bit of a loner and a contrarian, was raised as a diplomatic brat. Born in Germany, he was the son of a Canadian foreign-service officer based in Ottawa who took his family on overseas postings. Mark was schooled in England and Brazil, but mostly in Canada. He did two years at Carleton University in his hometown and then four more at the University of Victoria, where he double-majored in geography and strategic studies. It was there that WestJet's future vice-president of strategic planning ("and paranoia," he adds) got a grounding in areas that would be crucial in planning and developing an airline. As he told his alma mater's alumni magazine, a military-history course taught him tactics of combat campaigns. One geographer was "a no-nonsense professor who raised the bar on how to digest and process reams of information in an organized fashion." Another was a rebel in the department, a stance that appealed to Hill—as did the philosophy of a historian specializing in American politics and diplomacy who "encouraged my somewhat off-the-wall, argumentative, unconventional-wisdom thinking."

At UVic the quick-moving, sardonic student with curly brown hair and a penetrating gaze met the woman who would become his first wife. A field-hockey player for the university, she came from Calgary, where her father ran Western Canada's largest architectural firm, the Cohos Evamy Partners, and the oil boomtown is where the couple headed after Hill graduated in 1985. He soon immersed himself as a worker in Ron Ghitter's unsuccessful campaign against Don Getty for leadership of the Alberta Progressive Conservative Party to replace premier Peter Lougheed. Unlikely as it now seems, for the next three years Hill worked in the public sector as assistant to the president of the Alberta Vocational Centre, an adult-education school, who became his mentor. Fred Speckeen, who later became a Presbyterian minister, was a role model by demonstrating in his people skills "how to conduct yourself, personally and professionally. Although I'm not a people person," Hill notes. "That's why I don't have the big staff today, and I know it and everybody knows it. But I understood how he did it and why he did it—and why I could never do it." Speckeen also taught him how to handle a piece of paper only once, which is how his protegé manages to deal with the flood of information that crosses his desk at his home office in Victoria, where he now spends most of his work week.

Calgary was never his favourite city ("I can't stand the horrible winter climate," he says) and when government downsizing threatened his job at the vocational centre, he decided to leave town and earn an MBA at the University of British Columbia. On the first day of classes, he thought, *Oh, boy, have I made a mistake!* "The courses seemed more quantitative and I'm more of a qualitative guy. I dropped to half-time and left at the end of the first semester."

Returning reluctantly to Calgary in 1989, he worked in commercial real estate for Knowlton Realty and did a couple of small deals with Hanover Management. "Clive Beddoe was a big wheel in real estate and had the reputation of being a really sharp guy. Maybe a little tough, but at the end of the day pretty fair. A lot of guys would shake in their boots about doing a deal with him." (Hearing of his reputation today, Beddoe says, "I find that extraordinary. I didn't consider myself a tough negotiator; I thought I wasn't hard enough. I believe in trying to create a win-win situation. The only way you win in the long run is to show people how they can win also.")

One day in 1990, Hill—then in his late twenties—phoned Hanover about a possible tenant for a building, only to learn his contact there had left the company. Instead, Beddoe fielded the call. Mostly as a joke, Hill told him, "I've heard you've lost somebody. I'd better come down and put in my application."

"Why don't you do that?" Hanover needed somebody immediately to handle minor leasing matters; Hill still isn't sure today if Beddoe even knew who he was.

"I came in as asset manager, which was re-leasing a bunch of downtown real estate we had. Clive had a reputation then for being really moody. In hindsight, he had just gone through a divorce—and having gone through two myself, I know exactly why some days he could snap at people. So I was a bit nervous about that. But he gave me a lot of leeway. Clive is a master motivational guy. He understands what people's hot buttons are and how to set up a remuneration package that pushes those buttons."

Beddoe soon discovered one of Hill's buttons. Often at week's end they had a beer at The Chowder House, a restaurant owned by Clive's brother Steve, who had followed him to Calgary. Clive was pursuing his passion for flying as he earned his multi-engine, helicopter, and night ratings. One Friday afternoon in the fall of 1990, as they were discussing his flying, Hill—who'd never piloted any aircraft—said, "I've always wanted to do that."

"If you fill the Lougheed Building 97 per cent," Beddoe replied, "I'll buy you a private pilot's licence." The Lougheed was an eighty-year-old Class C office building at a time when similar properties had 40-per-cent vacancies. Today, he recalls Hill as "young, somewhat idealistic, and a very hard driver." Within six weeks his inspired employee had the building nearly full and by December was taking his first flying lessons—at Tim Morgan's school—and was soloing the following summer. Overseeing his mountain-flying course was Tara Linke, a young instructor who later became one of WestJet's women pilots.

Over the next few years, Beddoe gave Hill more opportunities to shine and increase his income annually. In one deal, the young go-getter wrote a business plan to acquire a warehouse building in Arizona that involved a Fanny Mae (a mortgage from the Federal National Mortgage Association in the United States) and several foreclosures that let Hanover buy at the bottom of the market and sell out with a nice profit.

Hill kept flying and became increasingly intrigued with the economics of the airline industry. One of his first public observations on the subject was that his boss's eight-seater Cessna was often flying half-empty on his trips to Vancouver. "Mark, who is unbelievably cheap and hates waste of any kind, thought it would be a good idea to sell seats on board the airplane," Clive Beddoe says. "And because we operated through Time Air and his commercial licences, we could do this. Mark proposed that we offer seats at half the cost of an Air Canada ticket."

And then on a rainy day in June 1994, Mark Hill wandered into Tim Morgan's office in the oldest hangar at the

29

Calgary airport. Unable to fly, waiting for the weather to clear, he began chatting about the airline business. He'd read about the great success story of this American low-cost carrier, Southwest Airlines, and told Morgan what he knew about it. Finally, he asked, "Why wouldn't it work here?"

"I don't see why it wouldn't work here." Morgan offered some of his thoughts on the subject and concluded: "Should we look at it?"

"If you give me some information, I'll build a model to say what this thing would cost." Hill began researching the subject at night and on weekends. "And then," he says, "Tim brought Don Bell in."

THE COMPUTER GUY: CARING FOR CUSTOMERS

Don Bell, a third-generation Calgarian, was the son of an electrical engineer who worked for TransAlta, a power-generation company. Don's first after-school jobs set the tone for his career. As a twelve-year-old delivering *The Calgary Albertan*, he was so service-oriented that he got a letter from one woman customer rhapsodizing about "the best service I have ever had from any *Albertan* boy, and that is covering the last 17 years, older than what you are." (His mother, whom he credits with teaching him how to treat people, has kept the testimonial.) Interested in anything with a motor, he became especially smitten with the aircraft business during his high-school years. His father chartered helicopters for power-line patrol from a flying service; the owner's widow became the wife of Gerry Stauffer, a local aviation legend known as the Chief, who had a company that overhauled airplane engines. Through the family connection, Don began there as a part-time mechanic's helper in 1972, aged sixteen. After high school, he moved to Medicine Hat to work with Stauffer's brother and nephew and got his pilot's licence at twenty and his commercial rating a year later.

Not wanting to be a professional flyer, and influenced by his dad, he took engineering at the University of Calgary, where he found his first career. After two years of courses, he shifted his interest to the more exciting new world of microprocessors. He answered a call from the university hospital's ambulatory-care centre, which needed help in developing a test programming project using microcomputers, predecessors to the personal computer—something he, like most other people, knew little about. This was the Stone Age. It was just a year after Ed Roberts, founder of the now-defunct New Mexican company MITS (Micro Instrumentation Telemetry Systems), designed the first desktop PC—the Altair, named for a *Star Trek* episode—and the very year Bill Gates and Paul Allen licensed their BASIC language to Roberts.

Inspired, Bell quit university and launched a software business with five friends as backers and about the same number of employees. His MITS Computer Systems, a private integrated computer systems supplier, assembled hardware from kits and began the hard slog of peddling machines with a mere 16 kilobytes of memory and a bulky 10-megabyte hard disk the size of a long filing cabinet. "They were completely unreliable, failed all the time, and the software was very buggy," he says. More promising was software development for accounting and the hotel, dental, and medical fields, and the resale of restaurant and labour-management programs. The company grew to several branches across Western Canada, an office in Toronto, and a hotel-software business in San Francisco, which he later sold. Two of the products were so successful that employees spun them off into separate companies, which are still flourishing. Clinicare serves North American medical group practices and Genie

31

Computer Systems develops and markets clinical and practice management software in Canada.

"Don is an interesting character—very unassuming,"
says Mark Mackenzie, who worked in service and sales at
MITS for eighteen years, helping open offices from Vancouver
to Toronto. "When he was selling, he did none of the text-
book things, but he was very successful. We had a pretty
heavy customer focus—a twenty-four-hour service operation. It was a typical small company, boom or bust, but most
of us were team players and were required to wear multiple
hats. Everybody had to be self-sufficient. Since leaving MITS,
I've begun to realize how much of a role Don played in my
life and my aspirations."

By the late 1980s, Bell was flush enough to buy out his
original investors. As president and chief salesman, manag-
ing up to forty employees, he says, "I learned how to run a
business, how to treat people—and customer service. It was
all about knowing your product, treating your people right,
treating the customers right, taking care of them." (He treated his employees well enough that several of them would
eventually join him at WestJet, including networks manager
Steve Stretch, who'd been a customer-service technician
with MITS.) "It was really tough to make a living at it. But
we survived and we were moderately successful. And I was
able to afford to buy a couple of airplanes that we used in
the business."

He had a Piper Comanche, a Cessna 310 and then a sleek
Cessna 340, which he flew across the west every week for
work and pleasure. With his airline transport rating, he even
took time off from his business occasionally to fill in as a
pilot on corporate jets and turboprops.

One of the people he could talk aviation with was Tim
Morgan. His folks and the parents of Morgan's wife were

friends. Bell sometimes called him for advice and they saw one another from time to time, usually at the airport. At one get-together in 1994, Bell showed him an article in *Fortune* magazine about Herb Kelleher, the head of Southwest Airlines. "You ever hear of this guy? You might be interested in this."

Oh, yes. "Come to a meeting we're having next Saturday," Morgan told Bell.

2

PREPARE FOR TAKEOFF
Doing due diligence

We've been telling corny jokes for 25 years and so far no one has gotten up and left. Having fun in flight has made us stand out since Day One. And nobody has played more games or sung more boarding procedures than our ten original flight attendants. Today, we'd like to say thanks for 25 years of entertaining captive audiences.

   —Southwest Airlines ad, celebrating its veteran
      employees in 1996

The cover of *Fortune* that Don Bell showed Tim Morgan featured the beaming face of Southwest's chair and a headline that shouted "Is Herb Kelleher America's Best CEO?" Judging by the article inside, he might well have been: "While household-name carriers like American, United, and Delta lost billions over the past four years, only Southwest among the big airlines remained consistently profitable. ('We didn't make much for a while there,' says Kelleher. 'It was like being the tallest guy in a tribe of dwarfs.')" By mid-1994, the no-frills airline had a market capitalization of nearly $5 billion and an attractive price/earnings multiple (the price of the stock divided by the earnings per share). Southwest had become the major carrier in the busiest US air travel markets. The magazine credited its success to Kelleher's open, caring, can-do style of leadership, quoting one industry analyst: "He is the sort of manager who will stay out with a mechanic in some bar until four o'clock in the morning to find out what is going on. And then he will fix whatever is wrong."

Morgan and Mark Hill devoured the details about Southwest's cost-saving reliance on only a single type of airplane (the Boeing 737), its startlingly low fares ($58 US for the average 600-kilometre flight), its lightning-fast turnarounds at the gate (twenty minutes, half the competitors' time). Then there was the extraordinarily enlightened work environment: model union-management relations, profit-sharing for employees, and the wielding of humour to entertain passengers and lighten the stresses of the job.

Early on they had meetings with various interested parties, including Don Bell and Morgan's brother, Darcy. Hill, in his weekend discussions with Tim Morgan, had come to understand some of the very basic financial elements of the airline industry. Rushing into Clive Beddoe's office one Monday morning, he said, "I've got an idea"—why not start a small regional airline?

"Why would we do that?" his boss recalls asking. "The other guys would gang up on us and wipe us out. So why would it work?"

Although Hill didn't yet quite know the answers, Beddoe says, "he was almost to the point of being speechless, he was so excited. He didn't know at that time the Canadian airline business had been deregulated in 1988. He didn't know anything." Curbing his skepticism, Beddoe agreed to let his employee drop all leasing duties and show him why a new carrier could fly in the troubled Canadian skies.

"If I'm going to do this, I haven't got time for real estate," Hill said.

They turned their receptionist into a leasing agent reporting to Hill.

"I'm going to need a computer."

Beddoe bought Hill a series of three increasingly powerful computers.

"One thing I've always found," Beddoe says now, "is that you can never underestimate the power of people's energy. Mark focused his energies, morning, noon, and night. It ultimately cost him his marriage. Tim introduced Don to Mark and the three of them got together and gathered data from different sources. Mark would occasionally come and tell me where they were going with it. And I would challenge it and tell him where it would fail. I was the nay-sayer. The obvious thing was that Air Canada and Canadian *would* gang up and kill us. But by that time he'd researched Southwest."

SOUTHWEST EXPOSURE

Southwest was only one of the airlines that Hill was studying that summer of 1994 as he and Morgan planned what they were then calling, for want of a better name, MorganAir. As for what kind of airline it would be, the universe was wide open—the sky was the limit. Morgan describes himself as "a person who wants a decision and wants it in black and white." He began researching decision-influencing facts such as aircraft types, airport fees, and licensing. Meantime, Hill was doing what he does best, thinking about strategy after dredging up voluminous data and synthesizing them into an understandable whole for presentation in a business plan. Because the Internet hadn't yet become a common research tool for the public—one that Hill would later play like a virtuoso on a violin—the self-described "flunky" spent days

37

holed up in a university library, reading books and photo-copying newspaper and magazine articles about the complex airline business.

The mythmaking accounts about WestJet suggest the founders examined the operations of twenty different air-lines around the world. Not true, Mark Hill says. He focused on just a handful of successful low-cost carriers in the United States while absorbing the overall state of the indus-try in Canada. Three US companies in particular fascinated him: Pacific Southwest Airlines, Southwest Airlines, and ValuJet.

In the years since, there have been at least six books written about Southwest, extolling it as the true pioneer among friendly, fun-oriented, low-cost carriers. But as Hill learned, the first such airline was *Pacific* Southwest Airlines (PSA), founded in San Diego in 1949 to serve California and eventually the neighbouring western states. Founder Kenny Friedkin, who never wore a suit, considered his staff an ex-tended family and insisted that everyone be on a first-name basis. As a result, employees pitched in during the early years to keep PSA aloft. Pilots loaded baggage, and baggage handlers and pilots helped mechanics put de-icing boots on the wings in bad weather. Passengers came to love an airline that adorned the noses of its airplanes with wide painted smiles (sparking the slogans "California is all smiles" and "Catch our smile to ... "). A former editor of the *San Francisco Chronicle* says, "Their business was to get you there feeling pretty good. Sometimes it was like a party, with a little band or guitar player. And they were on time and had good-looking girls." The young, miniskirted flight atten-dants were overtly friendly: "We treated customers as if they were guests in our homes," one veteran recalls. Andy Andrews, who became CEO after Friedkin's death, liked to say: "Life is tough enough. If we're not having fun, let's go do something else."

Although PSA was profitable throughout the 1960s, flying Boeing 727s and 737s, it faced a bitter strike and ascending fuel costs over the next decade. And by the mid-1980s, consolidation in the industry had created formidable competition on its home turf. Two years after American Airlines acquired Air California in 1986, USAir swallowed the final traces of PSA, burying its distinctive culture and removing all the smiles from its fifty-five aircraft.

But Pacific Southwest had long since inspired another low-cost carrier, which brassily borrowed part of its name and most of its style. Southwest Airlines was the creation of Rollin King, a Harvard Business School grad with a commercial pilot's licence and a tiny commuter airline in San Antonio, Texas. In 1966, King conceived a plan to fly the Golden Triangle of San Antonio, Houston, and Dallas, and approached his lawyer, Herb Kelleher, who weighed the competition and thought the man was bonkers. The client persevered; Kelleher invested in the venture and eventually became King's innovative and much-loved chair and chief executive officer.

But not before the first president, industry veteran Lamar Muse, set the tone for the start-up, which after legal challenges finally took off in 1971. Among other imaginative ploys, he ran an "Open Letter to Raquel Welch" asking her to be one of the airline's beautiful "hostesses." The flight attendants chosen from the deluge of applications wore red hot-pants and knee-high boots. He encouraged them to make Texas-style inflight announcements: "Soon as we get up in the air, we want you to kick off your shoes, loosen your tie, an' let us put a little love in your life on our way to Big D." They complemented the corporate slogan, "Now

Somebody Else Up There Loves You," originally inspired by the fact that the airline flew from Love Field in Dallas. When competing against Braniff International's predatory $13 (US) price point between Dallas and Houston, Muse took out a double-page ad accusing his rival of trying to shoot Southwest down and offering passengers a free bottle of good liquor if they flew his airline for its regular $26 fare.

On a more serious note, his philosophy of ceding power to his people to do whatever it took to get their jobs done helped spark innovation. His habit of questioning attendants after a flight led to improvements in customer service. And, perhaps most important, his offer of a generous profit-sharing plan for every eligible employee—the first in the North American industry—returned about 15 per cent of net profits to the staff. The company also fully matched employee contributions to corporate-run retirement plans. Those two monetary investments paid remarkable dividends in what Muse calls "the quality, enthusiasm, and esprit de corps of the Southwest Airlines group of employees." It also paid off for them more tangibly: "There is not a twenty-year employee of Southwest who is not a millionaire."

His people needed the camaraderie and incentives to keep motivated. The early years were fraught with legal battles as major airlines sought to down their low-cost, low-fare competitor. Colleen Barrett, the current president and chief operating officer, acknowledges, "The warrior mentality, the very fight to survive, is truly what created our culture." One Southwest warrior who didn't survive then was Lamar Muse. Despite his successes in a combative environment, he had to resign in 1978 after mounting a power play against Rollin King, the major stockholder.

Howard Putnam, the marketing vice-president, became president and chief executive officer for four years. He oversaw the writing of a fifty-two-word mission statement that he took with him everywhere, spreading its message to employees, customers, and investors: Southwest vowed to "provide safe and comfortable air transportation in commuter short-haul markets ... at prices competitive with automobiles and buses ..." When Putnam resigned in 1981 to run the ailing Braniff, Herb Kelleher took the controls and most of the credit for making Southwest such a financial and cultural achievement until retiring as CEO (but staying on as chair) in 2001. Like his predecessors, he encouraged the hiring of people who liked to have fun and who believed that, despite working harder and earning less than at other airlines, they would come out on top through profit-sharing. Which is how the culture of empowerment evolved: veteran employees remember that, from the start, few said "We can't do it" or "That's not my job."

During his research, Mark Hill began to understand how that culture set the airline apart. Jody Hoffer Gittell, a Brandeis University management professor and faculty member of the Massachussetts Institute of Technology's Global Airline Industry Program, has studied Southwest's management practices. In her mind, its "secret" is "its ability to build and sustain high-performance relationships among managers, employees, unions, and suppliers. These relationships are characterized by shared goals, shared knowledge, and mutual respect." Its success is not due to a non-union staff—Southwest is one of the most highly unionized airlines in the United States. Nor is it due to using only one type of aircraft or a simple point-to-point route system (while the majors use a large hub airport with radiating spokes)—the unsuccessful Continental Lite and United Shuttle have done that too. No, Gittell insists, it's the fact that Southwest hires and trains for what she calls relationship competence in such so-called soft skills as customer orientation and teamwork ability.

As Kelleher has explained, "Another airline can go out and get airplanes. They can acquire ticket-counter space at the terminal. They can buy baggage conveyors and tugs. But the hardest thing for a competitor to imitate—in the customer-service business, at least—is attitude. Esprit de corps is the way that you treat customers and the way that you feel about people. And it's very difficult to emulate that, because you can't do it mechanically, and you can't do it programmatically, and you can't do it according to a formula."

One of Kelleher's strengths was an open-door policy for the average employee. A vice-president once complained to him, "Herb, it's easier for a ramp agent, a flight attendant, a pilot, or a mechanic to get in to see you than it is for me."

"Let me explain the reason for that," Keller replied. "They're more important than you are."

The result, as Mark Hill noted in 1994, was that in the previous couple of years Southwest had won the US Department of Transportation's "triple crown" for best on-time performance, best baggage handling, and fewest customer complaints among all major carriers—honours that it would continue to earn.

Herb Kelleher, who understood the tenets of teamwork, was also bringing a discipline to Southwest. The CEO had the airline hew to a strictly delineated flight plan for growth and a game plan to achieve it. In *Nuts! Southwest Airlines' Crazy Recipe for Business and Personal Success*, management consultants Kevin and Jackie Freiberg write knowledgeably of their client: "Southwest is successful because it is willing to forgo revenue-generating opportunities in markets that would disproporitonately increase its costs. By focusing on profitability instead of market share, the company has demonstrated the discipline to do without market segments that don't fit within its niche. And no carrier knows its niche as well as Southwest."[1] (WestJet would later translate that philosophy into an admonition to potential markets: "Use us or lose us.")

Nuts! hadn't yet been published when the Canadians were doing their due diligence. If it had been, perhaps they would have taken its rhapsodic treatment of Southwest with several grains of salt—considering that its authors were paid consultants of the airline. But the founders were impressed at the time by Southwest's financial results and the admiring arms-length reports in the business press, including the *Fortune* article that Don Bell had pointed out. One of the most impressive facts was Southwest's stimulation effect on the marketplace. Inevitably, its entry into a viable market stimulated business, expanding it rather than simply stealing existing customers from competitors. Hill noted that when the airline started service between Chicago and Louisville, Kentucky, in 1993, only about eight thousand people were flying the route; within a year, the number tripled. The same effect happened when Southwest introduced a $49 (US) fare—one-fifth its major rival's—between St. Louis and Kansas. Often this surge of people involved the so-called Visiting Friends and Relatives (VFR) market, those who hadn't thought of flying these routes until it became competitive with driving or hopping a bus.

Hill was also encouraged by the more recent success of ValuJet, a no-frills airline that began flying scheduled routes out of Atlanta in 1993 and swiftly grew to serve more than twenty eastern US cities with fifty-one aircraft. Emulating Southwest in its casualness and low costs, while pioneering in areas such as paperless electronic ticketing, the start-up proved profitable almost from the first flight. The company had just gone public and its stock was soaring—prices would leap 500 per cent in less than a year. Its president, Lewis Jordan, was a graduate of Southern Airways, where as a senior executive he had to deal with an airplane crash in Georgia that killed seventy-two people, and of Continental Airlines, where he was forced to resign as COO while it lay mired in bankruptcy. He came to ValuJet at the behest of its co-founder, Robert Priddy, a former colleague at Southern. Jordan was a penny-pincher, even insisting that employees turn in their old pens and flashlight batteries before they could get new ones. The airline paid low wages and contracted out its maintenance. But it tried to copy Southwest's sense of fun: a cartoon critter appeared on all its airplanes, and Jordan characterized ValuJet as "the leisurely, family-oriented, affordable, fun type of airline." It all sounded a lot like Southwest, and ValuJet was well on its way to making a $20.7-million (US) profit in 1994.

At the time, there were no low-cost, low-fare scheduled airlines in Canada. And the traditional carriers were in trouble. In 1985, the industry was virtually deregulated as airlines could now have full and free access to all domestic markets. Brian Mulroney's Conservative government privatized Air Canada over the next few years, leaving the national flagship facing a dogfight with Canadian Airlines International Ltd. (CAIL) The subsequent deregulation of the industry prompted a flurry of mergers. Calgary-based CAIL itself was an unhappy fusion of Pacific Western Airlines, Canadian Pacific Air Lines, Eastern Provincial Airways, Wardair, Nordair, and Transair. Its subsidiary, Canadian Regional Airlines, was created from the regional commuter carriers Time Air, Ontario Express, and InterCanadian.

In 1992, Air Canada and CAIL had failed in an attempt to merge. As Wayne Skene reports in *Turbulence*, his book

44

about the deregulation of Canada's airline industry, "Both Canadian airlines were sadly uncompetitive. Air Canada's unit costs were a lumbering 18.2 cents per available seat mile [a standard way of measuring the costs of a flight] in 1991. CAIL's was slightly better at 14.6 cents. But neither compared with American Airlines' 11.0 cents. By 1992 the two Canadian airlines' unit costs combined were 50 per cent higher than the average unit cost for the seven largest US carriers. As both haemorrhaged with daily losses, Air Canada increased its seat capacity by fifteen per cent, forcing CAIL to increase its capacity as well, sparking a fare war that resulted in almost 70 per cent of passengers flying in southern Canada travelling below cost. This, in a business presumably bent on making profit."[2]

In the early 1990s, the first Gulf War had cut back international air travel and sent fuel prices soaring. Then a North American recession reduced domestic travel. By 1993, Air Canada and CAIL had collective net losses totalling nearly $2 billion. A year later, when Mark Hill was doing his research, American Airlines—although beleaguered by losses in its own country—had just invested $246 million in CAIL for one-third ownership. The deal saved CAIL but did nothing for the consumer flying either Canadian carrier. The result, as Skene writes, was "two, semi-crippled airlines telling the consumer, by virtue of their perilous financial circumstances, and with the tacit support of frightened and confused federal governments, that they could not afford to drop their prices. They could not even afford to go out of business."[3] In 1995, Canadian Airlines International would lose nearly $200 million.

With the two major airlines battling each other for passengers but unable to offer deep-discount fares, perhaps there was room for a little regional low-cost airline to slip in under their radar. In Tim Morgan's words, "No one had dared touch the airline business here because it was Air Canada and Canadian—two huge companies that everybody was scared to death of. But we could run between their toes and they could never quite step on us."

A MODEL AIRLINE

Mark Hill assimilated all the relevant data and supportive anecdotal accounts about the North American industry. He then freely stole from the experience of the low-cost carriers south of the border—the pioneering Pacific Southwest and its prospering successors, Southwest and ValuJet (neither of which he'd visited). Marrying all this to the pragmatic aviation research from Tim Morgan, he built a model of a low-cost airline. "And it included everything, right down to the number of peanuts people would consume," Beddoe says. Based on these data, Hill wrote their first business plan. It was a thirteen-page proposal for a modest carrier to serve a few cities in Alberta and British Columbia. Like the American discounters, it would tap those people who might otherwise drive a car or take a bus to their destinations or perhaps not even make the trip. The goal was "to build a new low-cost, high-margin airline operating initially between the Calgary and Vancouver International Airports. Our goal is not to take a piece of the existing market but to greatly expand the market. The number of passengers travelling between these two cities will be greatly increased due to the low fares. After the success of this route has been proven we will pick other destination pairs to add to the route system. These cities may include Victoria, Kelowna and Edmonton."

At the time, the proposed company was called ABC Air (for Alberta and British Columbia)—much too bland a name for an airline. Tim Morgan's brother, Darcy, sent a memo to the planners saying the name should be "easy to pronounce and remember; translate to nothing offensive in other languages, especially French; be suggestive of one of our primary goals IF AND ONLY IF it is universally accepted by the public." He'd gone through Cree and Blackfoot dictionaries, only to learn that appropriate words were too long, hard to pronounce, and changed in meaning depending on context—"Flying Eagle" could become "Man with flat rock." Any variation on "reliable," "timely," "safe," and "predictable," he wrote, sounded as bad as "ValuJet": "No customer can be proud of saying they flew with the budget airline. It's kind of like bragging about your Lada at a party ... just not done. Whereas we want to be understood as as inexpensive source of quality transportation ... we don't need to peg ourselves as low-budget—our competitors will take care of that for us." He did list some possibilities, among them FastJet, JetAir, TekJet, Trekker, Ascentia, and Altius (which he thought suggested "Alberta").

In the end, it was Mark Hill's mother, Moira, the diplomat's wife, who came up with an easily remembered name. Reviewing all the options—many with the word "jet" and some with "west" in them—she wondered, "Why don't you call it WestJet?"

"That's it," Clive Beddoe said. (So much for market analysis.)

A revised, and much more sophisticated, business plan reached Beddoe that fall. Enhanced over the next several months, plumped up with graphs and charts and a history of American low-cost carriers, it became the basis of the document that the founders would use to woo initial investors. What's surprising, nearly a decade later, is how much of it would survive the reality of running the actual airline. Among its key points was the creation of a casual, folksy style:

The public will know WestJet as the Fun Airline. A sense of humor will permeate everything we do.

Management will participate in all elements of the business. It will not be unusual to see management loading bags and pilots checking in passengers. Staff will be cross-trained where applicable.

Our philosophy is the Customer is truly #1 and is always right. We will go out of our way to prove it. Our corporate rule is there are no rules, other than fill seats!

In fact, it soon became clear to the founders that it was more important to put employees first in its list of priorities:

The airline industry is people intensive and WestJet's success will be based in major part on the skill with which we select, train and motivate our employees ...

WestJet will create a corporate culture that provides an environment of fun for both Customers and employees. We believe that happy employees and Customers will inevitably create positive financial results.

The business plan then went on to quote from a Southwest Airlines annual report that underlined the importance of profit-sharing and company-contributed stock to motivate employees—along with "the psychic satisfactions of pride, excitement, fun and collective fulfillment."

A summary pointed out that, as proven by low-cost regional airlines in the United States, "low fares stimulate demand, thereby dramatically increasing the size of the market. When people get used to traveling, they expand their horizons and stimulate the whole industry. WestJet fares will be, on average, 55% lower than those of the competition."

And that's what finally convinced Clive Beddoe—and subsequent investors—about the viability of a new regional airline: Mark Hill's well-documented research on the effect of market stimulation, which even included a book on the subject. "It turned my head 180 degrees from a nay-sayer when he demonstrated to me what happened when low fares were brought into the market ... Every market where this phenomenon had occurred, there had been an increase in traffic—a 40-per-cent reduction in fare on average produced a 140-per-cent increase in traffic. In some markets, it had produced 800 per cent. He also demonstrated that the biggest stimulation had been in the [routes where] people wanted to fly—no good having low fares between Seattle and Calgary if there is no interest in people to visit relatives, the VFR market. Once he had proven it, I absolutely believed in it."

The revised plan would prove prophetic. Although its initial aims were modest—three airplanes and four cities, with the addition of Winnipeg—the founders had lofty ambitions: "The advantage of Canada over the United States when proposing low cost, low fare service is that WestJet is similar to a Wal-Mart arriving in small town America. Once Wal-Mart arrives, the competition is boxed out, for ever. WestJet can very quickly capture the entire Canadian marketplace by offering service on selected high-profile routes."

3

FIND GOOD FLIGHT INSTRUCTORS
Mentors and advisors

On balance, ADD [attention deficit disorder] has been a positive. People with it tend to be more creative, to think outside the box more. They take more risks. I was at a conference of educators at Harvard, talking about ADD in my life, and one asked why I had never taken any medication for it. I said, "I'm afraid I'll take it once, blow a circuit, and then I'd be like the rest of you."
—Morris Air and JetBlue founder David Neeleman

We were probably ADD ourselves with so many ideas.
—WestJet co-founder Mark Hill

The meeting was a replay of what could have been the opening scene of *The Southwest Airlines Story*. When Clive Beddoe went to consult his lawyer about launching WestJet Airlines, he got the same negative reaction that Rollin King had received in telling his legal counsel, Herb Kelleher, that he wanted to start a little airline in Texas. Actually, Daryl Fridhandler, a senior corporate lawyer with the prestigious Calgary firm of Burnet, Duckworth & Palmer LLP (BD&P), had already heard the news through Beddoe's wife. "Clive's thinking of starting an airline," Ruth told Fridhandler's wife.

To which her husband replied: "There's no way. He'll lose his shirt." It was a dampening message he later delivered in person to Beddoe, who had delegated his Hanover Group real-estate affairs to Fridhandler's partner.

As the lawyer once told *Lexpert*, a Canadian legal-business magazine, he thought at the time: "Clive's a real estate guy, very successful, but conservative. He knows the business, he knows the numbers really well, so he's not going to throw money [away] ... [Y]ou see people starting airlines because it's a neat thing to do, and a lot of them don't do it right, and the money's gone."[1] But, as King did with Kelleher, Beddoe soon convinced Fridhandler of the dollars-and-cents sagacity of the low-cost concept and the lawyer wound up investing in the enterprise.

Both men knew that three of the foursome planning WestJet had no background in commercial aviation—and while Tim Morgan was a skilled professional pilot, his business experience was in running a flying school and a minor charter service, not an airline. On their own, the quartet would have little plausibility with potential investors. The only way Beddoe knew to gain believability was to buy it. "We had to go and find ourselves a person, a disenfranchised Southwest executive, a retiring president of a low-cost airline, somebody who had the credibility that we could borrow and [who could] endorse us so we could raise the capital," Beddoe says.

In late 1994, he was having just that conversation with Mark Hill in their Hanover office in the old *Calgary Herald* building downtown. Stepping into the reception area, they saw a young man on the sofa, waiting to present a business plan to Beddoe. Overhearing the two men talk, the visitor, a Mormon, mentioned that he'd just returned from his church's world headquarters in Salt Lake City, Utah. "You should talk to Morris Air. It's just been sold to Southwest Airlines. Maybe there's someone there who could help you."

Beddoe turned to Hill and said, "So there you are, Mark—there's your answer."

Southwest had acquired Morris Air for $129 million in stock and cash at the end of 1993. That was less than a decade after a twenty-five-year-old David Neeleman had co-founded the low-cost airline that focused on the Pacific Northwest. Neeleman, born in Brazil but raised in Salt Lake City, began his business career as a teenager helping run his grandfather's corner grocery store ("he taught me a lot about customer service, which to me means treating people nicely," the grandson says). Neeleman interrupted his studies at the University of Utah to be a Mormon missionary in the slums of São Paulo for two years. On returning home, he quit college in frustration, only later to be diagnosed as suffering from attention deficit disorder (ADD). Born-again as an entrepreneur, he packaged holiday condos and air charters to Hawaii, building up his agency into twenty employees and $8 million (US) in sales, until the airline he was using folded and his business went bankrupt. But his hard-sell approach had intrigued June Morris, the owner of a major local corporate travel agency. He collaborated with her on Morris Air, raising $20 million in venture capital and in one year almost doubled the company's value to $130 million. It grew from a charter-only to a scheduled carrier based on the no-frills model of Southwest, although with an even lower cost-per-mile structure. While Herb Kelleher was his idol, Neeleman brought his own innovations to Morris Air, including reservations agents who worked entirely from their homes and the first electronic ticketing system for airlines anywhere. The sale to Southwest netted him $22 million and a position there as executive vice-president.

After Clive Beddoe had suggested finding a refugee from Morris Air to inject credibility into WestJet, Mark Hill tracked down David Neeleman in Salt Lake City, where he was in transition winding up the last of his company's affairs. His secretary took a message and, on a chilly October day, Hill's pager rang in Earl's Restaurant in Calgary. This being 1994 BC (Before Cellphones in Alberta), he sprinted out to a telephone booth and spent the next hour or more in conversation with Neeleman, who expressed some interest in helping out WestJet.

Hill, Don Bell, Tim Morgan, and pilot John Takacs flew Beddoe's plane to Utah, where Neeleman was surrounded by boxes and tearful Morris employees upset about their family-like business breaking up. The Canadians were intrigued by his humble persona—his wallet was threadbare and a trench-coat he bought second-hand bore a cigarette burn. He drove them to lunch at a Mexican restaurant in his beat-up Mercedes. "He's ADD and all over the place," Hill remembers. "We were probably ADD ourselves with so many ideas." They showed him their business plan. Neeleman told them about the recent start-up of ValuJet, whose principals had consulted him, borrowed his idea of ticketless travel, and taken his advice to operate out of Atlanta. "Look at the multiples; these guys are making a fortune," he said.

As Hill tells the story, "We are in awe. We know nothing. This guy can add so much value. How do we tie this guy up in Canada so no one else can get him?" By this time, reports were surfacing about another Canadian low-cost carrier (Greyhound Air) in the planning stage. Neeleman would be a good fit with WestJet: "He was our age, he had good ideas, he'd done his missionary work in Brazil and I'd lived there, he liked technology and Don liked technology." Flying home, the heat in the recruiting team's airplane went off, but they were all boiling over with Neeleman's agreement to come to Canada, where he'd never been.

They brought him up for a three-day tour of Calgary and Vancouver to be presented to chamber of commerce and board of trade people, municipal politicians, and airport authorities who were now learning first-hand about WestJet's plans.

"We actually had a very good reception pretty much everywhere," says Clive Beddoe, who was impressed, if a little bemused, by Neeleman. "He's very much a visionary. Very much a numbers person"—less of a people person, he agrees. "He describes himself as being ADD: he can't sit focused on a point for very long. Very difficult to tie down. David was amazed at the cost of flying in Canada and that there was no low-cost carrier and he was also surprised to learn about the impediments to travel in Canada. The waterways, mountains, the geography, and climate." He came to realize that a motorist would have to spend a dozen hours driving from Calgary to Vancouver through the Rockies while a plane trip would take little more than an hour. As Neeleman imagined the potential Visiting Friends and Relatives market, Beddoe says, "he saw a huge opportunity. By the spring of 1995, we had him interested enough in the venture that I tried to convince him to become WestJet's president."

Neeleman's career with Southwest would end within a year as his aggressive personality rubbed Herb Kelleher and his people the wrong way ("Southwest is a great company," Neeleman explains, " but I'm an entrepreneur; I move at a different pace"). Yet although his termination contract with Southwest allowed him to work for a non-American carrier, he had no desire to run WestJet. He had nine children at home and was building a software company around a reservations system. Under his non-compete clause, he could refine the Morris Air e-ticketing system—turning it into the reservations program called Open Skies, which Hewlett-Packard later bought for a reported $22 million (US). (Don Bell had already realized that the system, even in embryonic form, was a valuable cost-saver, eliminating paper tickets, and had agreed to adapt it for WestJet's purposes at an initial cost of about $80,000.)

Beddoe did manage, however, manage to persuade him to join the board and invest $200,000 in the company. "You guys are smart," Hill recalls him saying. "You don't need me. I'll be in the background." While the founders had no problem investing in the company, none of them had considered becoming working executives. Given his background, and the fact that he and his wife had just started a family in 1993, Don Bell thought his role would be, should be, limited to choosing the corporate computer systems and getting them up and running. "I made it very public that I didn't want to work here; I didn't want to be in management," he says now. "I also wanted to be a pilot, I wanted to learn how to fly Boeings, and I was going to run my business a little bit and semi-retire." (He and Morgan did learn to fly 737s.) Beddoe himself was busy with his own ventures, especially the plastics company, and so was keen on hiring a high-profile president from the airline industry.

GUIDANCE COUNSELLORS

A mentor is generally defined as an experienced and trusted adviser. Although the word was inspired by a fatherly figure of that name in Homer's *The Odyssey*, business mentors don't have to wear a figurative toga as an all-knowing elder. Whatever their age, they can act like good flight instructors who by their experience and example, their coaching and encouragement, take you across distances and to altitudes you've never gone before. Having a mentor can demand humility, a swallowing of pride, an acknowledgement that you still have something to learn from someone who may be smarter, more skilled, or simply more seasoned. And not every potential role model works for everyone as a mentor in practice. David Neeleman wasn't one for Beddoe, who saw him more as a high-flying symbol of respectability—an airline guy with a good, relevant track record—to parade before people of influence and possible investors. "But David was essential," he admits. "Without him I wouldn't have gotten involved." Answering a brief questionnaire recently for the *Globe and Mail*, which asked who his mentors were, Beddoe responded, "I don't really have any. I'm a bit of a seat-of-the-pants guy, an opportunist who couldn't believe the amazing opportunity right in front of us." He sounded like the attention-deficient Neeleman in completing the statements, "My greatest weakness is" and "My greatest strength is" with the same phrase: " ... my lack of attention to detail."

The pragmatic Morgan, who found the thirty-five-year-old Neeleman flighty, says he did provide some inspiration and direction but little hands-on help ("He and I were oil and water"). Bell, however, knew Neeleman had both management skills and technological know-how to offer. And the young Mark Hill, more of a theorist and an airline strategist in the making, saw the unschooled but brilliant man from Morris Air as an inspiring source of specific information and overall wisdom in creating a low-cost carrier. Hill probably appealed to Neeleman because of his insightful and straight-out style of speaking (an astute WestJet employee would note much later, "Of our four no-bullshit founders, Mark's the most direct of all").

Over the next several months, Hill regularly flew down to Salt Lake City to get briefed, poring through the piles of business plans on Neeleman's desk for prospective airlines in the United States and Europe. "I studied all the price-stimulation modelling and reverse-engineered it to figure out how they did it," he says. "Every time we remodelled, the result was the same. It was very clear the economics were there. I did this until about spring of '95, when we started raising money."

Neeleman returned to Calgary to meet likely investors. One of them was acting if not as a mentor, then as a trusted adviser to Beddoe. Ron Greene, an old friend from sailing, was the founder of Renaissance Energy, a low-cost, high-profit petroleum company in Calgary that by aggressively exploiting its own oil and gas properties in the plains area of Alberta had built up annual sales of more than $500 million. (Within a few years, the innovative producer would agree to a $2.8-billion merger with Husky Oil.) Greene—a compact-looking, hard-riding recreational polo player—had a reputation of being as independent in his thinking as his company was in the petroleum industry. Peter C. Newman describes him in *Titans* as "a self-effacing, dedicated executive mentor and the quintessential shareholders' man. He has made a career of putting everyone else first."[2]

In late 1994, Greene got a call from Beddoe. "Back in the early '80s, Clive put together a group to go on a sailing trip and once a year the four of us went sailing. He and I saw each other socially but I had never done any business with him. He and Mark came into my office on a Friday and said he had a business plan they'd like to run by me and tell him whether it was crazy. Mark had a presentation all laid out on his laptop and a book about six inches thick with all his research about the discount-airline business. Clive left me with the binder and asked me to digest it as a friend." Greene's "sparse knowledge" of aviation came from his directorship a decade earlier with then federally owned

58

de Havilland of Canada. Now, although Greene had flown Southwest Airlines and superficially followed its success, the airline industry was not on his radar screen.

"I knew that Clive had some real business acumen and a lot of integrity. Those are the first two hurdles to looking at any business deal." Studying the proposal over the weekend, he found some attractive elements that reminded him of the fundamentals on which he'd built Renaissance Energy as a public company dependent on a cash-flow operation. "Clive had said they wanted to become the lowest-cost operator in the airline industry. They wanted to own their own assets to reduce one element of risk. Again, that was the Renaissance analogy because most of the other independents were leveraging themselves and leasing all their gas plants instead of owning them as we did. Clive wanted to finance with equity, not be highly leveraged as every other airline was."

Another key component of the WestJet business plan rang bells with Greene. Renaissance had a generous stock-purchase plan for all employees and a stock-option plan for pilots and executives. "We were on the same track: that people not just be employees but be shareholders and feel like part-owners of the business. At Renaissance these plans were the very foundation to attract the kind of employees who are motivated by the same things I'm motivated by; I have no doubt of that ... You want employees saying, 'Why don't we try this to save costs?'"

The following week, after more questions and comments, Greene surprised his friend by remarking, "Well, it looks so interesting to me, I'd like to be part of it."

Beddoe had no intention of asking him to invest. "Clive just wanted a checkup from the neck up; he wasn't looking for the chequebook," Mark Hill says. Ron Greene became the first outside investor and soon joined the board of directors, chairing the compensation committee that he knew would be so vital to WestJet's success. He would bring in some friends too, including Wilmot Mathews, an investment banker with Nesbitt Burns in Toronto. He'd been on the Renaissance board and became a long-term WestJet director as well as corralling other shareholders. Other than having their expenses covered, none of the directors then (as now) received any compensation or stock options. They were on the board as either significant shareholders or had other commitments to the company.

By the time Greene decided to invest, the founders had been hashing out their own financial involvements in the proposed venture. Beddoe's personal financing of just the research phase had climbed to roughly $150,000. At one meeting, he had to press the others on the level of contributions they were willing or able to make in seed capital. "You guys have to either put up the money or the whole thing falls apart," Tim Morgan quotes him as saying. Morgan eventually wrote a cheque for about $60,000. And while the other entrepreneur, Don Bell, had cash to inject, Hanover employee Mark Hill didn't, and had to settle for a smaller piece of the company in return for his sweat equity. Later, the four of them decided not to upset potential investors by allotting themselves too much cheap common stock at the start. Instead, they linked their percentages to the company's future success. When Beddoe told Daryl Fridhandler, "I need to be able to find a way to get these guys some shares," the lawyer introduced them to the concept of performance shares, a separate class of stock, which would convert to regular common shares depending on how well WestJet did. As Fridhandler says, "the performance shares were driven by actual, overall profitability," which could enrich the founders while benefiting the other investors as well. Beddoe—who decided he wouldn't accept a salary in assuming the role of CEO—later went out on a limb for his fellow founders and asked the board to extend the performance shares for another year so they could convert to a greater percentage of stock.

Fridhandler prepared an offering memorandum for the founders after they'd decided to make personal pitches, instead of using a corporate finance firm, to approach colleagues in the Calgary business community. Not everyone in town knew Beddoe. Richard Osler—who helped run Pemberton Securities in the city, and then opened his own investment firm while contributing to the *Financial Post*—says that for many, "Clive Beddoe came out of nowhere." But based on his reputation as a developer, Neeleman's American profile, and a refined business plan, the founding partners were asking each investor for a minimum subscription of $500,000 for shares priced at $1 apiece. Hill prepared a PowerPoint presentation with a couple of hundred slides on his laptop. It stressed a low-cost structure for a low-fare, short-haul western airline and the market-stimulation theory that would make it profitable. At one sales call, with Tim Morgan, an elderly executive of Fording Coal fell asleep during their presentation.

That was the only pitch that tanked. Within twenty-seven days, they had $8.45 million from only eleven external investors. Among them was Murph Hannon, head of Murcon Development, a private company with various interests, including oil and gas exploration and real-estate development, who would become a director. All of the founders brought in friends to invest. "I remember," Mark Hill says, "we were sitting with more than eight million bucks and saying, 'Holy shit, now we've really got to do it.'"

FLIGHTPATH

Beddoe, who himself put in a further $500,000, says one hot button for investors was "the data about what happened when you dropped fares by 40 or 50 per cent and the market would go up by 140 per cent." Another was creating a low-cost structure "by aligning the interests of the people with those of the company. We hadn't totally crystallized how we were going to do it—profit-sharing, stock—but we *were* going to do it." And a third was the description of real-world successes that some of the early investors got from David Neeleman and one of his colleagues, Dan Hersh, who had consulted to Morris Air and then helped found ValuJet.

Hersh was a tall, jovial veteran of Frontier Airlines of Denver, where he was vice-president of planning, and a boutique banking investment company in New York, where he was an aviation-securities analyst. In 1992, he became involved with ValuJet, buying 5 per cent of the start-up at founders' prices and preparing a strategic plan for an Atlanta-based scheduled airline. Earlier, he had received stock in Morris Air for advising Neeleman on route structure and strategy as the carrier switched to regular schedules from charters. His specialty was market stimulation. "When Morris Air cut prices, say 50 per cent, the Salt Lake–Oakland market could expect to double or triple if a Delta matched them. But Delta didn't match them and Morris Air got 80 per cent of that market." In the 1980s, Delta Air Lines averaged 12,000 passengers per quarter between the Utah and California centres. In 1995, two years after Morris began scheduling a tenth of Delta Air Lines' number of flights at half the fare, Morris had 50,000 passengers a quarter to Delta's 8,000. As Hersh pointed out, "Your worst case is if a competitor is matching you entirely." But, as he learned in coming to Canada to talk to the founders, that wasn't likely to happen with Air Canada versus WestJet. The operating

62

environment was abnormal as the major national airlines were losing money—and if they became too aggressive against a newcomer, they could be charged with predatory pricing (as Air Canada later was).

"The key to ValuJet, Morris, and WestJet," he says, "is that you could see that the big airlines, [operating] out of their hubs, were gouging their local markets in terms of price because there was limited or no choice." Not everything he suggested made sense. He didn't know local conditions, and his suggestion of a Calgary–Seattle route was shot down because the business communities in those cities had no substantial links. And when Hersh and Neeleman questioned the viability of a Calgary–Kelowna route, the founders had to explain that many Albertans had either moved to or holidayed in the BC Okanagan city.

Hersh impressed the original investors at a meeting where he explained graphically, on a chalkboard, what can happen when a discount airline enters a market. He drew a straight line depicting the road that motorists would take to get from point A to point B. Then above that would be three lines representing fare prices: Air Canada at the top, Canadian Airlines in the middle, and WestJet at the bottom. As Tim Morgan recounts Hersh's explanation, "You've got somebody in a minivan on the road, and when you bring the price of the ticket down, you can get him into your airplane. But Canadian and Air Canada see you do that and say 'Oh, can't let that happen, we'll do the same. We'll lower our prices.'" Now the minivan driver, and a whole lot of other motorists, decide to give up their vehicles to take the cheap flights. "Pretty soon, in theory, we get everybody off the road and into the airplanes. But the problem is the other guys' costs are so high, they can't sustain it. So they bump their prices up and the motorists fall right back down to flying WestJet. So it almost makes us recession-proof: no matter which way you do it, you'll get it coming up and you'll get it coming down."

Mark Hill would underline the point about the so-called rubber-tire market: "Five million people drive between Edmonton and Calgary every year. Our competitors are not Air Canada and Canadian. They're Ford and Toyota, and the guy selling gasoline on the side of the highway."

For Tim Morgan, Dan Hersh had more mentor-like qualities than David Neeleman. Interestingly, Hill points out that "Dan was very much of a mentor for Neeleman. He's very analytical whereas Dave thought in big, broad brushes." Hill got on well with Hersh, who says, "Mark was like a sponge, learning everything he could." For all the direction the wealthy co-founder of ValuJet gave the founders, he took no fee or shares. "I don't need money and I enjoy this. I taught Mark a lot and I did some work, but it was pretty simple." He has since advised his old friend Neeleman on the creation of JetBlue of New York; the low-cost airline flies larger planes with leather seats and satellite TV on longer routes than Southwest has done. Since its founding in 2000, JetBlue posts the best profit margin of any American carrier.

WestJet had chosen its mentors well.

SHAKING THE MONEY TREE

In September 1995, Daryl Fridhandler stepped into the office of a young new lawyer at Burnet, Duckworth & Palmer. He was getting acquainted with Dino DeLuca, fresh from a partnership at another firm where he had done work for Wardair and the creditors of a restructured Canadian Airlines. Spotting DeLuca's binders labelled with aviation topics, Fridhandler asked if he'd ever been involved in buying, leasing, or financing aircraft. Assured that DeLuca had, the senior partner closed the junior's door and confided: "We're going to have a new client that no one knows about and we need a lawyer to buy a couple of aircraft this fall."

Beddoe had decided they should purchase their own used jets rather than lease them, as Canadian Airlines was doing at a cost of about $250 million a year. To buy them, they needed more funds and this time they used an eastern brokerage house, Research Capital, to identify institutional investors to make minimum subscriptions of $100,000.

There was a certain urgency to their quest. Throughout 1995, as they were raising money and planning their airline, they tried to be as low-profile as a stealth bomber to prevent a possible rival from beating them to the marketplace. That spring, they had arranged for David Neeleman and Dan Hersh to meet Barry Lapointe, the president of Kelowna Flightcraft. The BC-based company flew a fleet of Boeing 727s to haul cargo for Purolator Courier and was interested in expanding into the passenger business with DC-9s. The WestJet principals and their advisers had a day-long meeting at the old Shell hangar at the Calgary airport to explore the possibilities. If WestJet could partner with Kelowna and piggyback on its existing domestic passenger licence, maybe the new airline could be operating that summer. But the prospect fizzled as fast as it had surfaced and Lapointe veered off in another direction.

As it turned out, he flew right into the arms of Greyhound Lines of Canada. The bus company, whose American parent had been suffering from the incursions of discount airlines, was anxious to launch an equivalent Canadian carrier. But only Canadian-owned airlines could get an operating licence and Dial Corporation of Phoenix owned two-thirds of Greyhound Canada. Kelowna, 100-percent Canadian, was well positioned as a partner. By the fall of the year, the Calgary office of Research Capital set up a meeting with Clive Beddoe, Mark Hill, David Neeleman, and Greyhound's president and CEO, Dick Huisman, a former senior vice-president of marketing at what was then Canadian Pacific Air Lines.

Hill picks up the story: "Dick is in the power position at the end of the boardroom table. He's very Dutch, very uptight. Clive never turns anybody off, but I see Dick as the enemy. Dick tells us about his business plans in general, without saying where Greyhound will fly. He's basically saying, 'You may as well quit.' My competitive juices are flowing. I knew he was trying to intimidate the shit out of us." At one point, Huisman mentioned that he'd consulted with Southwest Airlines and Morris Air. "David leans over and asks, 'Who were you talking to at Southwest?' And Dick couldn't name any names. David says, 'I'm on the executive planning committee there and I don't remember you. And who were you talking to about Morris Air because I used to run Morris Air?' You could see Dick shrink down in the chair, and if he'd had a tail it would be firmly lodged between the legs. That's when it got personal between WestJet and Greyhound Air."

Late in the year, Beddoe and Hill were in Toronto to meet possible backers. They woke up one morning in the Royal York Hotel to a report in the morning papers that Greyhound might really take to the skies. "We had five or six cancellations that day," Hill says. Yet they still had several sessions jammed into each day, including one with Frank Mersch, then a vice-president and director of Altamira Management. The high-flying, bearded fund manager strode late into his plush boardroom, put his hands on the polished table, and said, "Not another fucking airline!" Hearing highlights of their proposal, he insisted, "You can't fly an airplane from Calgary to Edmonton for $29."

"You want to make a bet?" Beddoe said.

By meeting's end, Hill says now, "We thought we'd got him. But they never did invest."

Research Capital's John Palumbo had particularly targeted the Ontario Teachers' Pension Plan, but Canada's second-largest pension fund (with assets of $40 billion) was having none of it. Teachers' Merchant Bank, its private-equity arm, simply wasn't interested in backing any airline, given the state of the industry, especially in its home province. "But John's persistent and Teachers' gave us half an hour," Beddoe recollects. "And a few hours later ... ."

It was the last day in town for the pair from WestJet and, riding the subway north to the fund's offices, they'd been pumped yet ready for rejection by the Teachers' team. "They were totally skeptical, arms folded," Hill says, "but then you could just see them getting into it. All of a sudden we've got three or four conversations going and they're saying, 'I want to see your modelling.'" As the afternoon wore on, he and Beddoe were glancing at their watches, delighted but still hopeful of making their evening flight. They returned the following week for further talks. As one of the bankers admitted afterwards, "We were extremely skeptical initially because most of the upstarts in Ontario went bankrupt pretty fast. But when we finally looked at WestJet's business plan, our opinion turned 180 degrees." The pension plan invested $10 million at $2\frac{1}{8}$ per share for a 24.2-per-cent stake. Among those persuaded was Brian Gibson, Teachers' senior vice-president of active equities, who served on the airline's board for five years. In all, WestJet's private placement with institutional investors totalled $28.5 million; at that point, the directors and senior officers controlled a quarter of the common shares. "It became very clear," Mark Hill says, "that we could not only pick our investors but pick the *quality* of our money."

The money from Eastern Canada—and the confidence it demonstrated for a Western-based carrier—would help make WestJet the most heavily capitalized airline start-up in North America in fifteen years. And if everything went according to plan, it expected to carry 300,000 passengers and generate $40 million in revenues during its maiden year. But first things first: the founders had to locate a couple of airplanes and a couple hundred employees.

4

BRING EVERYBODY ON BOARD
Profit-sharing and share-purchase plans

Hear you're starting an airline.
> —Kevin Jenkins, then CEO of Canadian Airlines, to his
> squash-partner friend and neighbour, Clive Beddoe

Just thinking about it.
> —Clive Beddoe, acting as nonchalantly as he can, while
> figuratively crossing his fingers behind his back and
> hoping his nose won't start growing longer

One of the many mantras that float through WestJet's world is "Hire for attitude, train for skill." It's an admirable appeal to have recruiters consider the power of the positive glass-half-full personality instead of simply weighing the talents and experience of a prospective employee. The idea is not merely to bring aboard all the people you need to get the job done but, more important, to bring people on board with a similar upbeat, inspiriting outlook on life and their livelihood. The concept has worked well for Southwest Airlines, whose president Herb Kelleher liked to say: "We'll train you on whatever it is you have to do, but the one thing Southwest cannot change in people is inherent attitudes."

This corporate incantation, however, comes with its own corollary: "But, you know, sometimes you really have to focus on skill first." That's a pretty obvious conclusion in the airline industry, where passenger safety demands well-trained pilots and mechanical engineers.

It was certainly true as WestJet's founders began hiring staff throughout 1995. Almost all of the original managers came from other airlines and not all of them had the right attitude or the long-term commitment to get a new carrier off the ground at any cost. Ideally, they might have modelled themselves on the four founders, who immersed themselves deeply in the start-up, sometimes to the point of obsession. Mark Hill's first marriage became one of the casualties: "I had also been working at Hanover and we had a new baby at the time and my wife was not particularly impressed." His self-imposed workload, he says, was a factor in their separation the following year. What incentive would it take to inspire employees to show a similar devotion—although not to the point of risking relationships?

The answer seemed obvious: a sense of real ownership. Turn them into true stakeholders. Then grant them the power to make their own decisions so they know they're making a genuine difference on the job. And the result will be, as Clive Beddoe has said, "a bunch of ordinary people doing extraordinary things."

In return for such dedication, WestJet's people were promised, almost from the start, that they would be rewarded directly. Rewarded for the physical and mental investments they made through their work. And rewarded for the financial contributions they could make in future to embryonic stock-purchase and profit-sharing plans.

In mid-'95, the staff was minuscule. Hill had now switched entirely to airline work, squeezed into a small office with a receptionist and the first inflight manager, Marlene Egland, who had done the same job at Time Air. They were in the old *Calgary Herald* building across the road from Canadian Airlines' downtown office; a brass street-level sign introduced the WestJet name quietly to the world. Hill, with his real-estate background, was handling sales and marketing by default. Egland was planning the recruitment of flight attendants. Tim Morgan had hired Gareth Davies, an expatriate Canadian, to be director of maintenance and to lead the search for airplanes. Davies, who oversaw maintenance at a Florida company that repaired and overhauled aircraft, happily packed up his station wagon and drove home with his family. A mechanic who'd worked with him earlier and while he was at WestJet describes him as a no-nonsense guy, "a very shrewd individual, but if he wanted to tell you where to go, he did."

By now, there was a new boss in place: that spring, Morgan had found their first COO in Glenn Pickard, whom he'd known as Time's chief operating officer. Pickard had been trying to help get another small airline off the ground. Unsuccessful, he came to WestJet, with perhaps less than full-blown enthusiasm. Pickard would become the first failure of the early hiring.

As COO, he was soon courting a marketing colleague from his Time Air days. Bill Lamberton was now at Canadian Airlines as product development manager for western Canada and California. Born in 1954, he'd grown up on a grain farm and in the dot-on-the-map town of Ormiston, Saskatchewan, which didn't get electricity and running water until he was fourteen. His entrepreneurial father became a local electrician and the postmaster, inspiring Bill to take a commerce degree at the University of Saskatchewan with a major in marketing and strategic planning. He then joined an independent travel agency and ended up part-owner as it grew into the largest in the province. But the airline business intrigued him and in 1981 he joined NorcanAir, a lifeline of Beavers and DC-3s for remote communities in the northern Saskatchewan bushland. As sales and marketing manager in Saskatoon, he helped develop commuter runs in small F-28 jets among cities in the south with connections to Winnipeg, Calgary, and Minneapolis. In 1987, the owner sold out to Time Air during a period of consolidation when Pacific Western Airlines and Canadian Airlines were merging. Marketing for Time, Lamberton visited communities that had lost PWA's jets and pitched them on the merits of more frequent flights with smaller Dash 8 turboprops.

Time Air, which became a feeder connector for Canadian—the "blue" airline (for its predominant corporate colour)—competed against AirBC, part of Air Canada—the "red" airline. Time's main focus then was called switch marketing. Speaking about that era to luncheon meetings later, Lamberton said, "If all the people at this table were flying red, the blue marketing teams worked on getting them to fly blue, and if all the people at this table were flying blue, the red team tried to get them to fly red. The rest of the tables in the room weren't flying—it was unaffordable, it was elitist, it was unobtainable. All the marketing strategies in those days were to switch people: 'We'll put more bells and whistles and rubber chicken on our airplane so that we can switch you.' There was little attention paid to try to stimulate the market."

Through the early 1990s, markets shrank and costs ballooned with the first Gulf War, the general sales tax (GST), and union gains. Fares went up as frequencies dropped. In '92, as Time merged with three other airlines to form

Canadian Regional, Lamberton joined the parent Canadian Airlines' marketing team, focusing on route development, new products, and pricing tactics in the west. Over the next three years, the carrier nose-dived with even heavier losses. He was ripe when Glenn Pickard called in August 1995 to say he was working with some guys intent on starting an airline and would Bill be interested in coming aboard?

The forty-one-year-old Lamberton—lean, with a heavily lined brow (but without the goatee he has since adopted)—met an intense Clive Beddoe. The inexperienced airline co-founder wanted to pump the strategically placed marketing man from Canadian for proprietary information. "Here's the plan," Beddoe said, explaining WestJet's approach. "How are you going to stop us?" The circumspect Lamberton, surprised he wasn't being interviewed as a potential employee, was loath to reveal too much. But he realized—from his own family background and his experience with western airlines—that the idea of wooing the Visiting Friends and Relatives (VFR) market was brilliant. Families were scattered across the west and the one almighty fact about this half of the country was the Rockies and the Coast Mountains: "It was the rocks. There were rocks in Western Canada and we had to get over those, there was inclement weather, and there were islands, and there were ferries and buses that you could compete against. So I knew there was pent-up demand."

Lamberton had later sessions with Pickard and with Tim Morgan and Gareth Davies (both of whom he knew from Time Air), and spent several late nights with Mark Hill analyzing his business plan. On the August day Lamberton became convinced WestJet could fly, he called Pickard to arrange a meeting that evening. As he recalls the conversation, Pickard said, "I got to tell you something. Everything's going good, we're still deciding on this or that airplane and where we should have the hangar. Oh, by the way, I'm leaving."

"Glenn, I was just gonna tell you I'm joining, what the hell do you mean you're leaving?"

"Oh, don't worry about it. The king of Bahamas has made me an offer that I can't refuse, to run Bahamasair. And I'd kind of like to go to the Bahamas this winter anyway—it's getting to be September. But don't worry, Bill. The guy that's gonna replace me, you know him very well, I can't tell you who it is right now, but you're gonna love him."

With a wife and three young children to support, Lamberton was less inclined to leave a secure paycheque for a leaderless new airline. Two days later, he saw an announcement at work that the chief accounting officer of Canadian Regional, Don Clark, had just resigned. Almost immediately, Clark phoned him to ask if he'd decided to take the job at WestJet—where, he now learned, Clark would be his boss. Lamberton signed on at the end of October and working closely with a local ad agency started planning an advertising campaign for an airline that the founders hoped would begin flying on a date chosen for its inherent drama: Valentine's Day, 1996. The first ads were scheduled for early November, but would run only if WestJet could actually track down its first three airplanes by that fast-approaching February 14.

David Neeleman, with all his connections, had failed to locate any aircraft for his new Canadian colleagues. Gareth Davies was delegated to actively search for some to lease. Tim Morgan—then otherwise engaged in battling the bureaucracy for an operating licence—now says, "Our business plan was all based on DC-9s, not 737s. They had lower maintenance costs and were good, reliable airplanes. But every time we went to lease one, ValuJet would be there ahead of us, handing cash out. So we changed our direction and went with 737s. As it turns out, the DC-9 would have been the wrong airplane for us because it will not fly into a place like Kelowna."

Although United Airlines had some Boeing 737-200s to lease, they couldn't be licensed in Canada for technical reasons. Leasing rates for others had climbed to $100,000 a month from $40,000. Because sale prices were increasing more slowly than lease rates, it started to make sense to buy the 120-passenger aircraft outright. Finally, Davies tracked down two twenty-year-old former Canadian Airlines 737s that had flown ever so briefly for Astoria, an airline aimed at business flyers in Toronto and Montreal that had missed its target. Launched at the end of May, it shut down 117 days later when its New York lessor seized the aircraft because of missed payments. Davies and Beddoe flew in to meet the lessor and appeal to his paternal side: please sell your airplanes to this nice young upstart airline in Canada. The man had been wooed by many other hopeful buyers and his asking price was a daunting $7 million (US) for the pair. As Beddoe agreed to what seemed an extortionate price, Davies was kicking him under the table. "Look," Beddoe explained later, "it's a rising market and we've got to have them." When a Canadian bank would lend them only $2.5 million, Beddoe persuaded the vendor to carry the balance until the closing of the subequent financing. (WestJet promptly switched its banking to Toronto-Dominion.) Davies later lucked into a third 737, which Air Canada had owned. Because it would have cost too much for an American airline to check it out for US registry, WestJet was able to buy a $2 million airplane for a bargain $900,000.

But the airplane scavenger hunt had taken much too much time. Bill Lamberton was forced to postpone the

November ad campaign because there was still no assurance the three aircraft would be in place—checked out, refitted, and painted—to take off on Valentine's Day.

THE PEOPLE CHASE

And now people had to be found to fly and maintain the airplanes. For most of 1995, Tim Morgan was still on the payroll of Canadian Regional as a pilot. After the airline heard rumours about his involvement with WestJet, he received a letter to cease and desist or be fired. He took the letter to his director of flight operations and said, "I'm going to write back to them saying 'I'm not going to cease and desist—in fact, I can own shares in anything I want, including Air Canada, and you can't tell me not to do that.' He said, 'Great; do it.' [The issue] didn't resurface until about November of that year, when I told him, 'Look, we're going to start this airline, there's going to be a conflict of interest, I'm going to quit.'

"But he knew I was going to hire some pilots from Canadian—in fact, I had fifteen guys picked out that I wanted to hire. 'Okay,' he said, 'I know these guys will want to leave anyway. You let me keep them on staff—they can take holidays and do your training—but I need them to fly for me for at least three weeks to be able to fill the void.'

"So our crews are working for Canadian Regional and getting paid by them—and I'm training them."

Seventeen of the first twenty-seven pilots came from the Canadian Airlines subsidiary. Bruce Flodstedt, with seventeen years' flying experience, was one of them. The first time Tim Morgan talked to him about the new airline, he thought, *Tim's out of his mind, he's lost it.* "Pilots are a very skeptical group; it's part of our training to triple-check," Flodstedt says. But after a couple of conversations with

Morgan about the business plan, he realized there was nothing to lose. "I wasn't happy at Canadian Regional," he reminisces. "Time Air was a great place to work, a lot like WestJet. A smaller airline where you felt part of the family. There was share-purchasing, profit-sharing. When Canadian took us over, a lot of the working conditions changed. It deteriorated to the point we went on strike one summer."

As part of the first crop of WestJet pilots, he trained in Denver during two weeks of ground school using a United Airlines full-flight simulator for the 737s. "It was a fairly easy transition for most of us. It's a well-designed airplane, similar to the F-28s, and very, very forgiving," says Flodstedt (who much later became director of flight operations). All of the pilots were trained and fully certified with head-spinning speed by the end of the first year.

"Our original chief pilot, Rupert Bent, was a very intense individual, a Jamaican who did a good job of getting us up and running," Flodstedt says. "But he left shortly after we started." It was another case of hiring for skill in those early days and not focusing on an employee's personal style and sense of teamwork. Bent came from Air Jamaica, which was then an Air Canada affiliate. Some of his colleagues found him a really cool guy with his "Easy, mon" accent. But when Bent, fresh from the plateaus of his Caribbean island, started stressing the importance of a mountain-flying course, other pilots had to tell him, "Rupert, we've all been doing this for years." WestJet veterans say he was well trained but very much an old-school aviator who believed in the gold-bar style of the major airlines and tended to exercise his authority as chieftain of the aircraft. Constantly rubbing Morgan the wrong way, he was gone within the first year. Major George Hawey, a former commander of the Canadian Snowbirds precision flying team, replaced him on the understanding that he would eventually

77

return to being a regular pilot, which he did. Another original, Dave Lowing—"Father Dave"—had 737 experience flying for Dome Petroleum and became a patient adviser on handling the airplane over the next several years.

Ben Atkins—tall, with the rugged good looks of the archetypal pilot—had also flown speedy jets for Dome in the oil patch and then his own slow turboprop while servicing uranium mines in northern Saskatchewan. He couldn't believe the liberty WestJet's founders were offering its first batch of captains and first officers to decide how their flight operations would work. The pilots felt free to create their own system because, like most of the people at WestJet, they wanted to take a calculated gamble on making it work without the strictures of the unions that had scarred many of them. "It was groundbreaking to be where you could design your own job. They were saying 'This is *your* airline.'" For a start, there would be no seniority system, in which the pilots with the least experience pull the worst shifts. "You'd hear guys from Canadian Regional saying, 'I haven't had a Christmas off for three years because I'm junior. Let's make it fair for everybody.'" Ernie Byl, the first pilot to upgrade to captain, suggested an auction-style system that awarded his colleagues an assigned number of points they could bank and then use to bid for their preferred vacation blocks of five days or fewer and for days off in their monthly schedules. Actual flying time would be assigned, not bid on. "It provides for flexibility and fairness through the pilot group," says Byl (now the flight-operations administrative manager), "and as a result, our newest pilot will have the same quality of life in his or her schedule as our most senior pilot." As Tim Morgan points out, "the airline industry is extremely seniority-driven; they live and die by their numbers. In my experience as an F-28 pilot [with Canadian Regional],

I was both number one on the list and the bottom of the list. At the bottom, I worked every weekend, worked every holiday, never had summer holidays."

The business model had already decided that pilots and flight crew would all be based in Calgary, rather than be spread out in various cities in the style of the major airlines. Having them in one location could help build morale. Less crucial, but important culturally, was the pilots' general agreement to dispense with the staid uniform of stripes and starch the major airlines demanded. "We thought leather jackets were pretty uptown—'Hey, we don't have to wear suits!'" Atkins remembers. "And no airline hats. We didn't even want to wear ties, but we did. It was nice and refreshing to be part of that decision. What do we need to portray here, what image? We're casual, but professionals."

The quickly developing culture at WestJet delighted Stu McLean, an escapee from Kelowna Flightcraft, where he'd been reluctantly flying freight. The silver-haired, florid-faced old pro had a career that included test-flying stretch Convairs in the Mojave Desert, iceberg reconnaissance off Eastern Canada, geological mapping in Australia, and surveillance off the coast of Russia. Frustrated with the management attitudes at Kelowna—"There's no jobs out there; if you don't like it, there's the door"—he'd helped organize a pilots' union. In 1995, after answering an ad for a director of flight operations at WestJet (a role Tim Morgan would assume), he met Beddoe and Pickard. "I saw the light at the end of the tunnel. These are real people. I sensed the idealism that could turn into reality. They were promoting the idea that the industry needed a change ... There were no airplanes, there was a shell of a company, but a whole lot of hope—everybody knew this would work." He was offered a captaincy and began training with the second group of pilots hired.

Some of the pilots were soon helping Michele Derry in finding contacts for hotels and taxis in other cities, even driving to Edmonton to inspect accommodation. Derry had been a flight attendant at Time Air and then a planner of crew schedules for Canadian Regional. Because WestJet was an unknown quantity, "there were no hotels that wanted us," she says. "I couldn't get good rates and they wanted money upfront."

WestJet's luck in landing employees who were skilled and experienced (if not always attitudinally correct) continued when it went after aircraft-maintenance engineers. As it happened, Canadian Airlines was relocating its overhaul facility from Calgary to Vancouver. Engineers with deep roots in the community and fears about the future of the beleaguered carrier were disinclined to move with it. Among them was Tom Woods, who'd done line maintenance and heavy overhauls for eight years. His wife worked at Ron Greene's Renaissance Energy, where she heard about an airline starting up in town. Woods and seven others sent their résumés to Gareth Davies with the message: "You're looking for maintenance guys and here we are." In the end, WestJet hired as many as eight of the Canadian engineers. "We came in with at least eight years' experience on Boeing 737-200s. We all knew each other very well. 'Bring on the broken airplanes; we'll fix them.'"

Rob Bowerman was one of the first five maintenance people working on the used 737s. He'd maintained aircraft in the High Arctic and then in Calgary for Dome and eventually Canadian Regional. With input from others, he prepared the first maintenance proposal for WestJet. Although he started unofficially in October, he didn't quit his regular job until January 1996 when the first airplanes arrived. "I had more 737 parts in my toolbox than WestJet owned at the time," Bowerman says. "We were securing parts from everywhere. And at first we were everybody: the washer, groomer, engine guy, radio guy. We never had coffee breaks; it was that intense." After about six weeks, the original crew asked if WestJet was hiring anyone else. Why, their bosses wanted to know. We'd like a day off, they replied.

One of the most intriguing early employees was Herb Spear, who was in his early seventies when Tim Morgan approached him to supervise flight operations. Spear had got Morgan his first job, polishing airplanes ("and he's never forgiven me for it"), and for more than half a century had flown Otters and DC-3s in the Arctic, sold aircraft, and overseen flight operations for four Calgary-based oil companies. He was doing flight ops for Corporate Express when Morgan invited him for a coffee. When Clive Beddoe learned Spear's age, he said, "We can use your experience and as long as you can handle it, go ahead." Although Michele Derry took over from him about a year later, Herb Spear moved on to set up an occupational health and safety program.

If the pilots and engineers came with the requisite airline skills, most of the other employees—flight attendants and reservations and customer-service agents—were hired for their attitudes. The hopefuls were replying to newspaper ads for jobs with an unidentified new company. Marlene Egland and two assistants were recruiting and training flight attendants. They scheduled personal interviews for those who made the cut in group sessions. Typical of those with no industry background was Maria Haswell, a young single mother who'd studied psychology at university and was now working at the front desk of a Calgary gymnasium while modelling on the side. After an initial phone interview with WestJet, she eagerly showed up a week too early for an interactive group session. Haswell had been asked to make and bring something creative; she demonstrated a plane with movable wings she and her daughter had made of construction paper. Given several choices of live performance, she skipped joke-telling in favour of singing and doing a jig. She also did a cabin-baggage announcement and recommended an inexpensive activity for child passengers. In suggesting blank paper and a cheap line of bendable toys, "I knew I had the job." Haswell, with a long tumble of black hair and a gamine face, was one of the first forty flight attendants.

An early recruiting ad called for "dynamic, high energy, team professionals to work in a non-conventional airline environment. A sense of humor and adventure as well as unyielding commitment to customer service and productivity are prerequisites for all positions." Barry Tawse, although himself a former district sales manager and base manager for Time Air and Canadian Regional, recalls, "We weren't hiring any ex-airline people: they bring—pardon the pun—too much baggage with them." Tawse, a roly-poly guy with a rollicking sense of humour, was handling customer relations and communications for a call centre that would field reservations. At a time when he says he couldn't even spell "PC," he was delegated to help get the centre's computers into place. WestJet had taken a small space in the two-storey McTavish Building that had served as a warehouse for Loomis Courier near the airport (it was so unsuited for offices that its faulty air-exchange made the employees groggy for much of the first year). Tawse wound up pulling wires through walls and assembling furniture when he wasn't helping interview potential WestJetters.

"We hired waiters, barmaids, hospitality people. The industry said these people weren't going to survive. But we were hiring customer-service people." In the first two weeks of cattle calls at a local hotel, the recruiters saw three thousand people in up to ten groups a day. Without ever revealing the WestJet name, they told the candidates that the Canadian airline industry needed a real shakeup: "We've all seen flight crews looking like morticians' nightmares, looking more into the mirror than at their passengers." A recruiting team of three observed telltale signs of body language (impatient glances at watches) and usually culled about half of a group for follow-up interviews. "We wanted those wide-eyed and amazed and smiling people who wanted to ask questions but didn't know what to ask." One-on-one sessions narrowed the field to fifty-two call-centre people and twenty customer-service agents to work at airport check-in counters. The new staff would be making much less than their major-airline counterparts, but there would be travel benefits and the tantalizing promise of someday earning a portion of WestJet's profits.

SPLITTING THE POT

Getting employees on board to share a corporation's aspirations and collective responsibilities—its philosophy and physical and mental demands—is seldom easy. Early on, WestJet found a way to kick-start the process, a mutually beneficial method that other companies in other industries are increasingly favouring. In retrospect, the decision seems like a no-brainer for the airline. Again, it had a model in Southwest, which was the first carrier in North America to offer profit-sharing to its staff—and without asking for any wage concessions. Southwest puts 15 per cent of its pre-tax operating income into this initiative. A quarter of each employee's profit shares is then invested in Southwest stock in a share-purchase program. As Kevin and Jackie Freiberg note in *Nuts!* "One reason Southwest people are so willing to step up and assume ownership for the success of the business is that they personally have a lot at stake."

Beddoe took a slightly different approach: WestJet would share profits with all its people without obligating them to place any portion of those gains in stock purchase. But anyone who wanted to invest in the company's shares could do so and the airline would match it in an equal amount up to 20 per cent. There were a couple of pragmatic reasons for such initiatives. One was that WestJet was not offering a pension plan. The other was based on the fact that it paid only 95 per cent of the median salaries in its industry. First-officer pilots started at only $40,000 (captains at $65,000); customer-service agents were then earning only $10 an hour. But by adding its profit-sharing and share-purchase plans to that base, WestJet hoped that a flight attendant, for instance, stood to make 133 per cent of the market median. And the original pilots had the opportunity to buy a block of shares as a sort of signing bonus—Ben Atkins, for example, had $8,000 worth—and they were given annual stock options (in an agreement that would be renegotiated after the company later went public and the stock began to soar). "Options for pilots were a fundamental part of the business plan—to get them to think like managers," Beddoe says.

Clive Beddoe muses on the founders' thinking that lay behind the two incentive plans: "How can you get the employee base to focus on making a profit for their own benefit? We don't have the cash upfront [to pay them high salaries]. Why don't we make it a function of the market, so the more we make, the more the employees get a share of it? ... Share the margin [the ratio of operating income to sales revenue]. If we make a 10-per-cent margin, we give 10 per cent of that margin to employees [prorated to salary]. Cap it at 20 per cent. So that what you produce is an S-curve with a minimum of 10 and a maximum of 20. Profit becomes an honoured word ... It's not me, it's not you, but all of us to try to create a collective team spirit ... We've got to change the mindset of people away from draining the pot to building the pot."

That was the philosophy behind profit-sharing. "But it's no good doing that just in the short term. If we create these sorts of margins and they make us money, then clearly we are going to create a shareholder value and therefore if [employees] own stock in the company, they've got a major investment that they've got to protect ... You've got the whole team pulling in the same direction. And when things get tough, they're going to fight that much harder to get us back on track."

He asked Ron Greene, the director heading a compensation committee, to find the most generous staff stock-purchase plan in Calgary. It turned out to be a program in which the company would match an employee's contributions up to 5 per cent of his or her salary. Beddoe decided to raise the limit to 20 per cent of any salary and match it dollar for dollar. That's why most of the pilots who were early investors own stock worth between $2 million and $3 million today and many of the first flight attendants have earned hundreds of thousands of dollars' worth of shares. Lynette Bryant, a thirty-year-old customer-service agent who was one of the WestJet originals, says, "When I moved to Calgary, I had $20 to my name. One of the proudest moments in my life was the day I could pay off my house if I wanted to. I sold it instead and bought myself a bigger, nicer house—eight minutes from the airport." Like their counterparts at Southwest, some employees invest their profit shares directly into the share-purchase plan—but of their own volition. And their investments can be sheltered in a Registered Retirement Savings Plan as they build their personal pension plans.

After WestJet moved into its cramped McTavish Building quarters, Stu McLean says, "the guys in the first training group were in the office, anywhere they could find space, sitting down with the [flight-procedure] books, trying to help the other guys get through them." During this period, as the pilots bonded like comrades-in-arms, one of them called McLean. "We think we'd like to get together and buy shares in the company; are you interested?" Steve Brown asked. "We're going to talk to the executive team." Even though the first airplanes hadn't arrived, most of the pilots were keen to participate financially. Their bosses were happy to oblige. "I ended up with 4,500 shares at $2⅛. It split twice since then and yesterday it was $20," McLean told me in the summer of 2003. (By 2004, it had reached $30.)

READY TO LAUNCH

Throughout 1995, the people of WestJet were loath to let it be known that they were on the verge of launching an airline. Given the looming presence of two major Canadian carriers and the real possibility of a low-fare competitor—Greyhound Air—secrecy was paramount. *Loose lips down airships.* Kevin Jenkins, the CEO of Canadian Airlines and a squash partner of Clive Beddoe, casually mentioned one day that he'd heard his friend might be starting an airline. "Just thinking about it," Beddoe replied offhandedly.

But others at Canadian had heard the rumblings about a new airline and one of them told her friend, Siobhan Vinish, who'd just graduated from the University of Calgary with a bachelor of commerce in marketing. The English-born, Saskatoon-reared Vinish had been a marketing coordinator for McDonald's Restaurants in Calgary and, while at university, assisted the media buyer at the local branch of MacLaren McCann West, the marketing communications agency. In return for a ticket to Italy, Siobhan (pronounced Sha-von) also worked on a brochure and support materials for a community-investment project for Canadian Airlines, where she had a glimpse of the corporate culture: "I went to a meeting one time with the communications people and a research department, but they didn't know any of the people that they were meeting. They were working on this project together, but they'd done everything electronically. I remember thinking, *Oh my God, how could you have an environment where you don't even know those colleagues' faces?*"

Still, the airline industry was alluring and one summer evening she was out for a drink with her friend from Canadian, who pointed out Mark Hill at a nearby table. Vinish approached him and said she'd heard about his start-up and was interested in applying for a job in marketing. He considered the bold young woman with long hair, high cheekbones, and a direct gaze, and politely suggested she send him her résumé. She not only mailed him her c.v. but also kept sending news clippings and anything else she could find about the industry. They met for a conversation, but she heard nothing from him for at least another three months while doing marketing for a land developer. Learning that the airline was closer to a launch, she tied a news release about herself to toy battery-run foam airplanes and sent them to Hill, Don Clark, and Bill Lamberton. The gimmick, and her persistence, intrigued Lamberton, who called her in for an interview and hired her as a public-relations specialist on New Year's Day, 1996.

The irony of her new position was that WestJet didn't want anyone to know anything about it for another month. The airline still had only a licence to carry cargo; the passenger operating certificate wouldn't arrive until January 31. Meanwhile, the first two WestJet aircraft were being refitted and repainted in hangars of Second World War vintage at an industrial air centre in the hot and dry climate of Roswell, New Mexico. (A 737 takes about 200 litres of paint, weighing about 135 kilograms.) The straightforward colours originally chosen were what Bill Lamberton, raised on Saskatchewan football, calls "Roughrider green-and-white." His ad agency had wanted something more creative and, at the last minute, he and Mark Hill looked at various colours to replace the green and liked one that matched Lamberton's parka. The agency translated the hue as a cool teal.

Chief pilot Rupert Bent flew one of the remodelled airplanes to Great Falls, Montana—keeping it under wraps near the Alberta border—where the new flight attendants could practise emergency-evacuation measures on the real thing. Ten of them flew down on Beddoe's Cessna 421 and Don Bell's Cessna 340, while some engineers drove to meet them there. It was mid-winter and there was no hangar available. "It was like 25 below zero—a typical Great Falls day—and the wind's howling," Tim Morgan remembers. "And here comes our airplane with the paint job on it. They're all crying and all so excited. We have to do the training on the slides outside and they're freezing their butts off. But everybody's having a real good time."

By then, the founders had decided to announce their existence formally on February 5, when the call centre would open as a $1-million, six-city advertising campaign was launched for an airline that would begin flying at month's end. It was a leap year and, having lost the fun of Valentine's Day, they figured February 29 was a dramatic replacement. ("Our start-up employees all thought it was a management plot so that we would never, ever have to give away a ten-year pin," Lamberton says.)

Mark Hill had flown to Van Nuys, California, to ride in a chase plane taking beauty shots of WestJet's first aircraft for the media. The 737 flew into Calgary under cover of night on February 4 and was immediately swaddled up in a hangar while some employees wept with excitement. *My God*, thought Rosanna ("Ro") Imbrogno, a reservations trainer who'd cobbled together the call-centre agents' first schedules with a ruler on long sheets of paper, *we're starting an airline!*

The next morning, reporters arrived for the unveiling of the jet with its gleaming white body, and bold blue-and-teal nameplate and tail markings. They met the founders, employees, investors, and David Neeleman—"to kind of give the press the feeling that somebody had done this before," Lamberton says—and saw the reality behind the rumours about Canada's first low-cost, low-fare scheduled airline.

"WestJet intends to put the fun back into travel," Clive Beddoe told the massed media, "with more casual flight attendants, more casual uniforms, exceptional service, and a great smile on the face of every member of the WestJet team." As Neeleman explained to a TV reporter, "Check 'em in with a smile and we fly 'em up there and give 'em their bag—it's a very simple operation."

"New airline arrives with cheap fares," the *Calgary Herald* headlined as its lead story. "Air travellers in Western Canada are in for a low-fare bonanza with the introduction later this month of no-frills carrier WestJet Airlines Ltd. based in Calgary.

"For as little as $29 plus tax, Calgarians will be able to hop on a WestJet Boeing 737 for a one-way trip to Edmonton. A one-way flight to Vancouver will cost $59 ...

"Canadian Airlines International has signalled it will match the bargain-basement prices. Air Canada hasn't made a decision yet." A CBC television report quoted a spokesman for Canadian: "Pricing promotions are not uncommon. All carriers, including Canadian Airlines, will match to remain competitive."

The plucky little upstart had just joined the air battle and was about to take on the Red Baron, the Blue Devil—and worse.

5

FOLLOW YOUR OWN FLIGHT PLAN
Controlling destiny during early disasters

A WestJet pilot is making an announcement to his passengers:
"Sorry about the delay, folks—unfortunately we sucked a flock
of Ottawa bureacrats into our engines and it will take a few
minutes to clean up the mess."
—*Calgary Herald* cartoon, 1996, on Tim Morgan's
office wall, where it can be seen by Transport
Canada officials

It was an inauspicious debut. WestJet's founders and a large
flock of their 220 employees were crammed into the call centre at 6 a.m. on February 5, 1996, waiting for Barry Tawse to
do a countdown and push a button to open the phone lines for
the first time. Lynette Bryant, employee #44, a thirty-two-year-
old who'd worked at a travel agency, had done the
rudimentary three days' training for reservations agents, learning the airport codes and doing dry runs with her colleagues
on a closed, untested system. Within minutes of the launch,
she nervously fielded the first call. Surrounded by photogra-
phers, she laboriously led the nice woman on the other end of
the line through all the possibilities for her trip—but in the
end, after a lengthy conversation, Bryant didn't make the sale.

Fortunately, by then the phones were erupting like the lines at a Jerry Lewis telethon (and Lynette Bryant did land a customer). Tawse had told the agents, "If anyone has a problem when we go live, stand up or raise your hand." Soon, he says, it was like a nice Saskatchewan spring day "with gophers popping up from their holes. Everything they'd learned had disappeared. It was just nerves." Things became so busy that supervisors asked flight attendants only halfway through their training if they'd volunteer to answer calls.

"It was a scary day," Don Bell says now. Neeleman had licensed WestJet a spinoff of a ticketless reservations computer system designed for Morris Air by programmer Dave Evans of Salt Lake City—who hired programmer Greg McDaniel of Dallas to come to Calgary and write the WestJet system in three months. It was the first carrier in North America to use this enhanced Open Skies program. "We didn't really know if the computer system was going to work. We didn't know if the telephone system would work. It was all brand new, nothing off the shelf. There was no training program, no manual. We just kind of did it."

Media coverage of the launch and the low-fare ad campaign continued to spur reservations. At various times, the call centre had to commandeer backup staff from the executive ranks, including Mark Hill, his wife, and her sister; Don Bell, his wife, and his brother; and Bill Lamberton's wife. "We ran out of space within two months and spilled over into the boardroom, with computers ringing the table," says Ro Imbrogno, who was then a reservations supervisor.

The new Visiting Friends and Relatives market was being seduced by WestJet's service to Calgary, Edmonton, Winnipeg, Vancouver, Victoria, and Kelowna at year-round off-peak fares up to 70 per cent below the major airlines' usual prices. Peak fares—generally on Thursdays, Fridays, and Sundays—were at least 50-per-cent lower than their competitors' ($298 from Edmonton to Vancouver return compared to $656 on Air Canada). There were no first-class or business fares, only economy. A third plane would be added in early March when the schedule increased to 152 flights a week. As a news release from Edmonton Airports summed up WestJet's advantages:

> The company has the ability to offer these low fares due to the adoption of some basic cost-control principles. First, WestJet is financed by equity, which allows them to own rather than lease their three Boeing 737 aircraft. Second, WestJet is 100-percent ticketless with its own state-of-the-art computerized reservation system. Third, they operate only one type of aircraft, the Boeing 737 jet. This substanially reduces their costs of holding multiple parts inventories, crewing, and the need for training their employees on numerous aircraft types. Finally, WestJet's staff of 200 motivated, productive, and service-oriented people will be rewarded through the company's profit-sharing and stock plan, in addition to their salaries.
>
> The principal services that a low-cost carrier does not traditionally provide [are] the connection of baggage between one airline and another and the provision of meals. By not providing these elements to their short-distance travelers, WestJet has the ability to pass savings on to their customers through low fares.

At first, there were few bookings from business people (a clientele that generated three-quarters of Canadian Airlines' revenues). As predicted, the low fares lured many first-time flyers, including the odd grandparents who broke into tears on the phone at the prospect of being able to afford family visits. The novices needed serious hand-holding about this thing called air travel. Agents compiled an informal Top 10 list of their questions, including "Do your pilots have licences?"; "Okay, where's the airport?"; "Do you land at the airport?" (passengers would arrive, suitcases in hand, at WestJet's McTavish Building); and "Do we have to bring our own chairs?" (one senior wanted to know if she should take her lawn chair and some passengers actually did show up with seats or stools). The absence of paper ticketing also caused concern: "Where do I go to get my tickets?" An agent explained, "We're ticketless." The customer retorted, "How are you going to know me? I need a piece of paper." Passengers got confirmation numbers and these, plus driver's licences, were all they needed at check-in. The many who insisted on something more received trip itineraries by mail at first, then by fax, and eventually by e-mail. At least there were paper boarding passes: Don Bell designed them himself on his laptop and printed them out on a $600 printer borrowed from his own business.

On February 29, following a wheels-up party the night before, the first WestJet 737s took wing from the Calgary airport. This too was a less-than-auspicious start. The founders waited expectantly for their chief pilot to show up at the scheduled time. Rupert Bent didn't appear—for five minutes, ten minutes. He arrived at the loading bridge twelve minutes late. "And why was he late? Because he was briefing his crew," a still-upset Tim Morgan says today. "This is a big, huge, long briefing—the Air Canada way of doing things. And you don't think Clive wasn't spinning, wound up a little tight, and me too? Very first flight!" Pilot Stu McLean's day-long inaugural flights between Calgary, Edmonton, and Vancouver and back remain a blur: "Somehow we made it work with calculators and pencils." Ben Atkins says that during the next several months, "You'd come out of the cockpit and people getting off thanked you profusely."

Teresa DeMare, who'd been a manager at Calgary's Palliser Hotel, was frenetically training more agents to book reservations and serve customers at the airports. WestJet was operating with a corporate average of 68 employees per airplane compared with Air Canada's 157 and Canadian's 205. The initial flights had maintenance engineers on board; "we all flew with the airplanes in the beginning, almost like a fourth crew member," says Rob Bowerman. "We knew more about the 737's system than the pilots did." Often the mechanics helped clean the cabins, even greeted passengers. In those early days, the pilots might assist the maintenance crews in return—holding a transmission in place, for instance, while a mechanic wielded the wrench.

Serving the 120 passengers aboard each 737 were three flight attendants. The first crop had a mere ten days' training under supervisors Marcie Elliott and Glendora Beaver. "Only a handful had worked for airlines before; the rest of us literally winged it," Maria Haswell says. "They filled our heads with airport codes and safety. We learned the most dangerous times are the first five minutes of flying and the last eight minutes before landing." Compared to other airlines' programs, Haswell points out, the initial training was, um, "a crash course—we did not do one bit of customer service. We were told to be ourselves, have fun, and keep it clean." Elliott approved of Haswell's personal preference to begin on-board announcements with "Hi, everyone!" rather than "Ladies and gentlemen," and use her own spontaneous humour rather than canned jokes. Female flight attendants were positive about everything but their uniforms. The multicoloured shirts looked as if they'd been made from cast-off saris and were so obviously designed for small Asian women that they often ripped across the shoulders. They were later replaced with smart blue sweaters over white shirts with button-down collars and a discreet logo over the breast pocket (customer-service agents wear denim shirts). Although light bomber jackets were spiffy, the wind whistled through them and, because they bore no logos, passengers tended to ask if the attendants were on a skating or a cheerleading team.

The crews soon had a test of WestJet's core belief that its employees come first—look after them and they'll look after the passengers—and the corollary that the customer isn't always right. It happened on the watch of Carrie Winston, a tall blonde woman who'd applied for a flight-attendant's job by sending "the boldest, cheesiest cover letter I've ever written." In her interview, the recruiters tried to draw out more of her personality by asking how she'd handle a child running up and down the aisle of an airplane. "If it was the last straw, I'd hide behind a row of seats, put my arm out, and clothesline him," she joked. Her main interviewer, who had airline experience, paused and then confessed, "You have no idea how many times we've wanted to clothesline little Johnny." Soon after graduating with the first training group, Winston faced exactly that situation. When the seat-belt sign came on during a flight, a little boy kept running up and down the aisles and the mother refused to keep him in his seat. Not only that, she began screaming and swearing at Winston and smacked her in the nose. In desperation, the attendant grabbed the kid and tossed him into the seat. As they deplaned, the parents warned her that they knew Clive Beddoe and she was *so* fired. When other passengers gave her their business cards in a show of support, she ran crying into the cockpit. Winston feared for her job when protocol demanded she fill out an incident form.

"But my supervisor came to me and said even if these people did know Clive, they were never flying with us again.
They [WestJet] backed me up."

Lesley Hefferton, a Newfoundlander who joined the airline later, was in the middle of a takeoff announcement as
the jet was taxiing when she noticed some men passing a
large Coke bottle of liquor back and forth. She told pilot
Brian Spears, who emerged from the flight deck and said,
"This is my airplane; you're out of here." As the offenders
exited, other passengers applauded. That became a not-un-
typical corporate response to unruly customers (those who
would not be graced with the usual name of "guests").
Today, the attendants take comfort from a "flight safety
promise" signed by Clive Beddoe: "Safety is WestJet's core
value because we care deeply about the health and safety of
our fellow employees and our guests ... No disciplinary action will ever come from reporting a safety issue."

There were other safety issues during those salad days,
such as a loading bridge bumping into one airplane, a fork-
lift plowing into the bulkhead of another, and the following
night an aircraft washer hitting an electrical box with a high-
pressure hose—shorting out the entire hangar and disabling
some of WestJet's computer system. The incidents bothered
a stickler like pilot Bruce Flodstedt, who can chuckle now as
he says, "Everything here was new—and it was an extremely challenging environment ... The experience level needed
development. Not all the gate agents knew how to position
the bridge properly. We'd sit there for twenty minutes. We
were constantly bugging the fuellers for fuel. The flight
attendants were young and inexperienced. We didn't want
to import somebody else's problems, which is great in theory but you sacrifice some experience. It's okay trying to
build a fun, casual atmosphere, but safety is paramount.

Early on, some of the original group didn't realize how critical it was."

Meanwhile, Barry Tawse remembers, the company was suffering another series of "natural, coincidental, or intentional disasters." Within the first couple of weeks, an airplane broke down. Because WestJet had no ties to other airlines, scheduler Michele Derry, with no computer program to guide her, had to desperately reroute the two other airplanes "on paper and in my head" and buy tickets for some customers on other airlines. In another case, a pop can got sucked into an engine, which required a quarter-million-dollar replacement. One day the McTavish Building filled with smoke and, while evacuating the offices, Tawse observed the executives' priorities: Don Clark fled with his computer, Clive Beddoe took his books and files, and Mark Hill walked out with a model of a Boeing 737 and a photo of David Neeleman's family—and promptly slipped on the ice. The smoke had come from a seized bearing; there were no flames. But there was a real fire later that affected WestJet when a building in Salt Lake City that housed the Open Skies reservations system, which the airline used, burned to the ground in a case of suspected arson. *What the hell is happening here?* Mark Hill was wondering. *I don't believe this is coincidence.* "There was a fair bit of paranoia," Tawse says. "We had our building swept for bugs—any kind of devices."

GROUNDING THE COMPETITION

The paranoia was understandable. The new kid on the block had stirred up a hornet's nest in the big guys' backyards. Within a week of WestJet's unveiling its first airplane, Canadian Airlines matched the upstart's fares in ads with the theme "No-frills fares. With all the frills." Air Canada also decided to cut its prices. WestJet, even before it was flying, responded to their counterattack with its own campaign: "Imitation is the sincerest form of flattery ... Gosh, thanks guys. WestJet brought western Canadians the lowest airfares in a long time. Apparently the other guys thought it was a good idea too. After 20 years."

Combined with the majors' assault was the threat of the other low-fare airline that could crowd the air lanes and cloud WestJet's future. The day the Canadian ads appeared, Greyhound Air announced it would start its own economy air service on May 22—across Canada—with Kelowna Flightcraft and its six Boeing 727-200s, just acquired with a bank loan. Dick Huisman of Greyhound Canada had made good on his vow to compete with WestJet. The idea was to have the bus company's passengers combine road and air travel on a single ticket. With Winnipeg as a hub, two flights a day would link the major western cities with Hamilton, Toronto, and Ottawa.

If WestJet thought its ads were cheeky, they had nothing on the newcomer's. One TV commercial featured a greyhound walking up to the front tires of a competitor's airplane, lifting its leg, and peeing—with the message "We're marking new territory." Greyhound's mere existence was certain to pee off its rivals. Both WestJet and later Canadian filed formal complaints to the Canadian Transportation Agency that the new airline was using a "back-door" approach and unfairly avoiding federal regulations that require domestic carriers to be 75-per-cent owned or effectively controlled by Canadians (WestJet had about 5-per-cent foreign ownership). Dial Corp. of Phoenix owned 68.5 per cent of Greyhound Canada, the airline's profitable parent company, which argued that it was chartering the air service of the domestically licensed Kelowna Flightcraft and handling only its marketing.

The transportation agency didn't buy the argument. Until Greyhound got a domestic operating licence, its wings were clipped: "If the operation of the proposed air services commences, the agency will take all actions within its jurisdiction to prevent such operation, including the issuance—if necessary—of a cease-and-desist order against Greyhound." While returning money to about 13,000 people who'd already booked flights, the company scrambled to restructure. Dial took over the tour-bus business of Greyhound Canada, which became sole owner of the intercity bus line and part owner with Kelowna of the airline. On July 8, after an appeal to the federal Liberal cabinet, Greyhound Air finally launched. But there were more problems: the aircraft didn't have any corporate logos for weeks, it raised advertised prices substantially for flights booked after June 22, and perhaps most importantly, passengers soon learned that every cross-country flight had to stop in Winnipeg because the 727s were short-range aircraft. Unlike WestJet, the airline had also browned off travel agents by deciding not to deal with them (a strategy it later had to reverse). The month before Greyhound Air took to the skies, *Canadian Business* said it "has already secured the distinction of having executed one of the worst product launches in Canadian history."

Siobhan Vinish, WestJet's public-relations manager, came to recognize that the rival company was a different animal "because they didn't have a culture ... They were trying to copy, not re-invent, in a lot of ways. They just copied everything we did, price-wise ... And they were selling airline tickets in bus stations, they weren't dealing with travel agents, and were trying to do everything themselves."

By September 1997, Greyhound went to the great dog kennel in the sky. It was a victim of its relatively high costs (a big start-up debt to service and gas-guzzling thirty-year-old planes that required a third crew member to fly them); a misguided business strategy; and the loss of potentially lucrative summer revenue when its takeoff was delayed in part by the pro-active posture of the scrappy people at WestJet. That same month, Vistajet, an Ontario-based carrier with two leased 737s serving five cities, including Calgary, folded its wings too after less than four months.

GROUNDING YOURSELF

Delaying Greyhound's launch had been a delicious win for the westerners. The lesson to be learned was easy to comprehend, but frequently hard to make happen: follow your own flight path. In other words, control your destiny. And, as the ads in old comic books used to say, don't let the bullies on the beach kick sand in your face any more. This stance demands intelligent analysis, the courage to take action, and very often an additional burst of aggressive boldness summed up by the Yiddish word "chutzpah" (sometimes illustrated by the nerve of a man who murders his parents and then throws himself on the mercy of the court because he's an orphan).

WestJet was soon challenged to display the same sort of chutzpah in very different—and potentially disastrous—circumstances. Since its own launch, the airline had been embraced by the west, flying at nearly two-thirds capacity as it created a fresh market with its low fares. The majors were competing aggressively, forcing their little rival to match undercutting prices ($98 from Edmonton to Vancouver, for example). Yet WestJet had surpassed its target and within half a year carried 388,000 passengers, earned a profit of $2.5 million, and had $10 million in the bank. It was a better flying start than even Southwest Airlines had managed.

But by September 1996, with the approach of the first annual meeting for shareholders—including employees who held stock—storm clouds were visible on WestJet's horizon. On May 11, a ValuJet DC-9 had crashed into the Florida Everglades and all 110 people aboard died. In June, in the wake of media and political pressure, the Federal Aviation Administration shut down the airline. Eventually its maintenance contractor was convicted on a criminal charge and insurance companies had to pay victims' families tens of millions of dollars. It took three months for ValuJet to resume flying on a limited basis (and within two years it would merge with untainted AirTran Airways). The crash sent an abrupt chill through the ranks of North American airline regulators, who started viewing low-cost airlines with deep suspicion. This sense of distrust may have been a contributing factor in an out-of-left-field decision that Transport Canada made later that summer.

Before WestJet launched, Transport's local officials had approved an aircraft maintenance program that the ill-fated Astoria used and that WestJet acquired along with the two former Canadian Airlines 737s (although by now there were three other jets in the fleet also operating on the same maintenance manual). As Tim Morgan recalls, "I said to myself, *You know this is kind of unusual—we're starting off with what I consider is a mature maintenance program, yet we're not a mature airline, we're just brand new.* But we checked with Transport here in Calgary, [and it had] no issues—no problem. We don't have to do the checks and don't have to change the parts quite so often. I should have known better, but anyway, I didn't. Transport is in our office every single day, looking at the operation, they're all smiles, everything's going great. Then we do our first audit, and out from the woodwork come some people from the airline inspection division in Ottawa to start looking at our airplanes ... These inspectors started tearing these airplanes apart and realized that we were operating under this approved maintenance program (approved in Quebec, by the way) and that they didn't like us doing that."

One Wednesday morning in early September, unaware of how serious the situation had become, Clive Beddoe was on his way to meet an auditor from Transport Canada in the WestJet headquarters. With the company lot full, he parked his car across the street and was walking back to the building when a man stopped him. "I've got some advice for you," he said. "Run—don't walk—run and get yourself the best damn lawyer you can find."

When Beddoe expressed surprise, the man said, "I'm with Transport Canada and I've just come from a meeting and they have a decision in Ottawa. They want your ass."

"Who's they?"

"It started with Air Canada."

"What are you talking about?" Beddoe asked.

"They think you're another ValuJet. Get a lawyer—fast."

"What's your name?"

"This conversation never took place," the man said, striding away.

Beddoe walked into the meeting to learn that Transport Canada was taking the position that the maintenance program its local officials had signed off on should never have been allowed. WestJet should have been following the standard maintenance planning document (MPD) designed by Boeing Corporation. And now the government agency would be auditing the airline as if it had been—which essentially meant failing the company's airplanes.

"How can we possibly pass an audit on a document we've never used?" Beddoe asked the auditor.

"We're just doing our job," he quotes the man as saying.

"Have you ever audited a 737 operation before?" Beddoe wanted to know, his usually reserved voice starting to lose its cool.

"No."

"How much jet experience do you have?"

"None."

"How many engineers have you got on your team with 737 endorsement?"

"Well, none."

"What the fucking hell do you think you're doing here?" WestJet's CEO exploded. "You're not even qualified to be here."

"Don't you get mad at me," the auditor replied. "I'm not qualified and I know it."

Beddoe called the man's boss in Calgary and repeated the warning that the unidentified fellow had delivered in the parking lot. The official said there was no truth to the story, but listened as Beddoe suggested a solution to WestJet's problem. The airline would make the required transition to the Boeing maintenance program, but wanted ninety days to do it. There was a precedent: if it had sold one of the airplanes to Air Canada, that airline could operate it for three months during the transition to another program. Why not give WestJet the same grace period? Sounds reasonable, the local Transport Canada official said, agreeing they might have a deal.

But then *his* superior called Beddoe to say WestJet couldn't have the ninety days—a position he held until hearing the argument first-hand and recognizing then that it did sound reasonable, after all. Eventually, the discussion reached the deputy-minister level in Ottawa, where Beddoe and airline lawyers had been lobbying. David Anderson, the transport minister, wouldn't meet with them.

When someone leaked the news to the media that Transport Canada was actively auditing the airplanes, reporters interpreted this as a possible safety issue. A Boeing spokesman tried to squelch that inference, pointing out that a memo from the federal agency portraying the WestJet aircraft as dangerously old was simply wrong, based on bad data. WestJet's legal counsel, Daryl Fridhandler, has since said that "speculation was rife throughout that period. WestJet's non-union. There were some union people on the audit team. There were maybe some personality conflicts. Who knows?"

Meanwhile, when each of the four aircraft was being inspected, it had to be replaced in the schedule. Bill Lamberton, who oversaw route scheduling as well as marketing, says that for a couple of days WestJet had been renting substitute airplanes, small Beech 19s and Dash 7s, to fill in for the big 737s. "As days went by, we realized we were damaging ourselves a lot. And we were sitting around one afternoon and I'm saying, 'I don't think we can do this any more, guys, we can't be changing the schedule—we're losing credibility out there. Something's got to change.'"

Something did. It was September 16 and at 10 p.m.—the night before the airline's inaugural shareholders' meeting—Beddoe got the deputy minister on the phone. Once more he laid out his argument and again received reassurance: "Yes, you've got a deal. Let me just confirm it and I'll call you back."

"I've got to know tonight."

"I'll call you back in half an hour."

An hour later, there was still no call. Beddoe phoned the deputy, who was in bed. Denying that he'd agreed to phone back, he insisted, "I never said anything of the sort. You have no deal with us. The audit is proceeding."

Transport Canada was about to issue a notice that WestJet's operating certificate would be suspended if it failed to address major "deficiencies" within fifteen days, and Transport Minister Anderson was about to announce that his people were acting "out of an abundance of caution."

Shortly after midnight, the four founders and their executive team, including Don Clark, Gareth Davies, Bill Lamberton, and Siobhan Vinish, gathered around a board-room table for a crisis meeting. Beddoe had been thinking, *Well, let's change this whole thing around.* "The only way we can deal with this is to shut the airline down ourselves," he said. "So we have got to get into a control position—we must move from the defensive to the offensive and shut down the airline." Then WestJet would make the transition to the other maintenance program on its own terms, in its own time.

As drastic as that solution sounded, the others agreed it had a defensible logic: the airline could present the situation to its employees, outside investors, and the public with its own positive spin. They also decided that no WestJet people would be laid off during the shutdown ("If we'd even considered laying them off," Tim Morgan knew then, "that would have been the end of the culture of the airline.") Vinish immediately phoned the *Calgary Herald*, stopping the presses. Bill Lamberton began planning a campaign with his local advertising agency to run full-page ads that gave the airline's side of the story. ("Certainly from my point of view," he recollects, "I was scared shitless that we would never come out of it.") Beddoe then told key managers "We're pulling the pin tonight" and they began phoning their staff in the middle of the night. Michele Derry assured her pilots that the disruption shouldn't last more than three or four days and they'd remain on the payroll.

The next morning, Beddoe called his board of directors together to ask: "Do we press on or bail out—just sell it? We've got to make decisions. I need your support if we go on." To a man, they told him to persevere, knowing that in only half a year the airline had turned a $2.5-million profit. Ron Greene remembers, "The response from all of us was that we'd already proved the business plan and if it took more capital to get through the crisis, we'd do what it took."

Then the embattled CEO spoke before a previously scheduled meeting of shareholders, including the employee participants, to explain the technical reasons behind the voluntary grounding. "When Clive took the microphone and said what we had to do," Ron Greene says, "he was near tears. That sticks with you—it's tearing him apart and we're not going to let that happen." Vinish says: "There was a point where he was emotional and I put my hand on his arm to give him a moment. He had gone through two weeks of absolute hell with Transport Canada, jumping through every hoop. At that point, he was exhausted." At a hotel press conference that day, Beddoe delivered an angry statement about the situation. Vinish had arranged that, when reporters started asking hardball questions, WestJet controller Tanis Neely would take him by the arm and lead him to a waiting car—which she did, carving her boss a pathway through the crowd.

The biggest public-relations blunder was a decision not to have any staff at the airports first thing that morning to avoid subjecting them to the anticipated verbal abuse. "It was the right motivation—wrong decision," Beddoe says now, reflecting on how irate and bewildered passengers felt at the empty check-in counters. Realizing the error, the executive team dispatched customer-service agents to the airports that afternoon.

Frustrated customers who called in during the first twenty-four hours for an explanation heard a taped message saying WestJet didn't know exactly when it would be in the air again. "When we turned on the phones again," Tawse says, "there were heated moments where people went ballistic." The first customers shouted, "I'll never fly with you again" and "I'll have my lawyer call you." But then a surprising sympathy began to develop, an understanding of WestJet's plight, and a venting of westerners' traditional antipathy towards Air Canada and the federal government. Leaving their names and phone numbers, the callers asked reservation agents to contact them when WestJet was airborne again.

That support was quickly reflected in the media. One *Herald* story was headlined "Stranded passengers point finger at Ottawa" and began: "Stranded WestJet Airlines passengers, faced with travel delays and the uncertainty of being shunted to standby status on other airlines' flights this week, shrugged off safety concerns and leaned toward blaming government for their inconvenience." WestJetters' T-shirts quickly sported a cartoon showing an airplane in chains, attached to a stake being pounded in by a Transport Canada bureaucrat. The backlash prompted a spokesman for the transport minister to go on the defensive, admitting that "it's not as if [the inspectors] found a glaring, imminent safety problem" while denying any suggestion that Transport Canada's action was meant to prop up the ailing Canadian Airlines.

Meanwhile, the reaction of employees was exceeding the founders' rosiest expectations. "Within a couple of hours of high anxiety setting in," Tawse recalls, "it was *Braveheart* all over again—people wanted to paint their faces blue and teal and march with Clive." In fact, WestJetters did demonstrate with banners along busy McKnight Boulevard near their headquarters and the resulting traffic jam of curious motorists, honking their horns in support, led to a three-car pileup. While Tawse himself wrote more than two thousand non-form letters to inconvenienced passengers, the phones were staffed by flight crew and executives to back up reservations agents deluged with incoming calls. During the first three days, stranded flyers learned they could claim reimbursement from WestJet, even if their replacement tickets on other airlines cost as much as $1,000. There were stories of pilots buying passengers tickets out of their own pockets. In one case, the accounting department immediately couriered a refund to an Edmonton man who had to re-book a flight to Vancouver for medical treatment.

Flight attendants were doubling as junior accountants and file clerks. Idled pilots attended trade shows with WestJet salespeople and some of them helped the maintenance engineers, who were shouldering the heaviest burden. Their monumental task was to check every part of each airplane, replace any that didn't conform to the new maintenance planning document's specifications. Gutting the airplanes, they laid out all the mechanical innards on the hangar floor and the tarmac. (A 737 has about 357,000 parts, with about 600,000 bolts and rivets fastening them together.) One morning Beddoe arrived, took a deep breath, and said, "You know where all this stuff goes?"

"It was the busiest time in my life," says Tom Woods, who by then was a maintenance inspector. "Seven days a week, averaging eighteen hours a day." Often the mechanics were at the hangar twenty-four hours a day, catching naps on the floor. "Wives would bring us suppers. Executives would bring Chinese food. Pilots would sweep the hangar floor." Despite the stresses of the work, "it was fun. Nobody ever said it can't be done. Just roll up your sleeves." There was a certain irony to the exercise: "Any items that had to be brought up to MPD standard were ordered and replaced. The reliability of the aircraft deteriorated because we put on unknown parts rather than the proven parts we had. A new overhaul part would fail on install and the one you took off worked just fine. We had parts that would fail after twenty hours; eventually, we weeded them out."

To audit all the historical records of the airplanes, WestJet called on Steve Ogle, an American mechanic with Learjet who later became a quality-assurance and maintenance director and an aircraft acquisitor for a major US Department of Defense contractor. He was greeted in Calgary by three of the Canadian airline's founders without knowing they were the owners (and he didn't learn any different until the following year). He found them resolutely low-key and casual: "Nobody else had a tie on and I was the only one wearing a suit. That night I bought the first pair of blue jeans I owned. I haven't put my suit on since." Ogle educated a team of reservations operators, customer-service agents, and even a flight attendant to help him create a presentable operating history of the 737s out of 180 boxes of records.

As time went on, things literally got a little hairy. Carrie Winston threw a Transport Sucks Party in her crammed one-room apartment with almost all the pilots and other flight attendants letting off steam by sporting outlandish hairstyles. Tim Morgan, Don Clark, and several managers had shaved their heads in a fit of morale-boosting. A couple of days before planes were scheduled to fly again, some female employees arrived at a meeting with maintenance director Gareth Davies wearing pull-on bald headpieces. Bill Lamberton set up Billy's Bar in a commissary in the McTavish headquarters and he and Mark Hill invited employees in to let their hair down and have a drink.

For the accounting department, the challenge was to maintain relationships with the new airline's skeptical suppliers. The corporate controller was Sandy Campbell, a thirty-six-year-old certified general accountant, raised on Prince Edward Island in an air-force family. He reported to chief operating officer/financial officer Don Clark, who'd hired Campbell at Time Air and was his boss at Canadian Regional Airlines. "It was a nightmare," Campbell says. "I started May 13, 1996. ValuJet crashed into the Everglades May 11. I thought, *Oh, my God, what have I got myself into?* My first day at work, I met with Revenue Canada. The GST returns hadn't been filed. There wasn't a month-end close. The balance sheet prepared for the board didn't even balance; it was out by $90,000. Revenue Canada wanted to know where their money was." (One of his staff, Jan Kersteins, had joined WestJet as an experienced accountant in March 1996: "Within the first week, I went for lunch and came back to a greasy old aircraft-parts box—full of thousands of dollars. I had no idea who had put it there. They really didn't know what to do.") Having straightened all that out, Campbell was now faced with reassuring suppliers and the bankers who administered WestJet's Visa and MasterCard accounts. "Don actually told the CIBC to leave the building. He lost it." The airline switched its Visa to the TD Bank. "You never forget your friends. We still have MasterCard with Alberta Treasury, which asked what they could do to help us."

For Hobe Horton, the shutdown offered breathing room. A high-school teacher who'd run an innovative sideline business compiling computerized statistics for provincial track and field meets, he was a friend of Don Bell. After a sabbatical studying computer information systems on a fellowship at DeVry Calgary, the private institute of technology, he asked Bell for a job. He'd come aboard in February and was soon sorting out a rash of statistics, including data on employee productivity at the embryonic call centre. "We made so many mistakes," he says. "So many of us knew nothing, so it was kind of cool." He found the shutdown an unexpected chance to retrench and design better systems. "It was a blessing in disguise, a time to get all our data processing started up again so we knew what we were supposed to get done during the day. Before, we would say, 'Wonder what's going to happen today?'"

It took seventeen sweat- and stress-filled days, but WestJet's people completed everything—from replacement of 120 different components to exhaustive documentation—to the satisfaction of Transport Canada's inspectors. Strictly as a sop to the feds, two top maintenance officials were moved to other positions, neither of them demotions. Gareth Davies was bumped up to vice-president, technical services, and Bill Kwas became director of quality control. All the pilots signed an open letter to the maintenance department, thanking the engineers for their work and reaffirming their confidence in the safety of the aircraft. Reservations agents cracked open a magnum of champagne that had been sitting in the call centre since the shutdown began. "At the end of the day," Clive Beddoe says, "we're a stronger company because of what happened." Director Ron Greene says, "It was very positive in instilling the culture in everybody." Mechanic Rob Bowerman says simply, "It was a defining moment"—a phrase that pops up when other veteran WestJet employees describe the voluntary suspension.

On October 4, the airplanes were off the ground again, 70-per-cent full, and with Beddoe and his family aboard the first flight. Three days earlier, Bill Lamberton had ads in newspapers across the west offering fares of only $49 to wherever

WestJet flew. A later ad showing a historic photo of a pioneer family in front of a sod-roofed hut had the tag line: "Where would we all be if they had given up?" The text noted, "We've all been through a lot these past few weeks, but as the saying goes ... through adversity comes strength." Another, headlined "You can't keep a good airline down," pictured a 737 in the air labelled "This is an actual WestJet plane in full flight" and "You'll note the ground is 32,000 feet below."

Lamberton had been more nervous about the relaunch than he had been about the original start-up six months before. Early on the morning the first $49 ads appeared, he sat in the call centre waiting for the phones to light up. They did, growing in volume as readers went through their newspapers across the west. "By 9 o'clock—no word of a lie—I knew that we would be okay *forever*. I knew that people believed in us." Although the grounding had cost the company $5 million in labour and lost fares, revenues soared after the resumption as westerners rallied 'round their home-grown airline.

While the passengers loved WestJet, staff at the other airlines didn't. Dale Tinevez, an Albertan who'd been a station manager for Time Air and Canadian Regional, came to the new carrier as manager of airports in late 1996 just after the grounding (replacing the first manager, Dan Gallagher from Canadian Regional, who'd hired him). Tinevez has memories of company cars being defaced with broken eggs and toilet paper. Pilot Ben Atkins says some resentful Air Canada employees left baggage trucks in the way of WestJet aircraft: "They'd go for coffee and it was a big drama to get them to move them." Flight attendant Carrie Winston recalls "ramp agents and baggage handlers with stickers saying 'WestJet' in a red circle with a line through it like a no-smoking sign. It was very uncomfortable."

Competitors had cause to worry. At the end of its first calendar year, WestJet had flown about 760,000 passengers and the workforce had doubled to 450. And it managed to finish its first fiscal year, reflecting ten months of operation, with a nominal profit on revenue of $37 million. Although the results didn't warrant it, employees in the profit-sharing plan received $500 apiece. It was a token of the airline's gratitude for their roles in rescuing it from the federal bureaucrats—and who knows what other malign forces out there plotting its downfall.

TAKE 'EM FOR A RIDE

In WestJet's annals, the next most telling example of controlling its own destiny occurred when the company had to take on American private enterprise. In doing so, the airline followed a flight path it hoped was well under the radar of aviation and law-enforcement authorities in the United States. Steve Ogle, having overseen the historical documentation of airplanes during the grounding, returned to WestJet under contract for the following five years. He helped acquire aircraft, oversaw quality assurance, and ultimately came on staff as chief maintenance officer. In early 1998, Gareth Davies asked him to help do due diligence on a twenty-year-old 737 being overhauled in Santa Barbara, California. In April, WestJet bought the airplane, but Boeing officials found that the company doing the modifications, Santa Barbara Aerospace, had skipped some crucial steps. Structural work, as well as a new paint job and reconfiguring of the interior, were supposed to be done by June 1. Ogle says Santa Barbara used inexperienced mechanics who damaged the skin of the aircraft and other components over the next two months. WestJet finally ordered them to stop and brought in a Calgary aircraft contractor and Boeing experts to redo the structural tasks. Santa Barbara was only allowed to repaint and refit the interior—some parts of which it managed to lose in its hangar for a week.

Late that year, Gareth Davies told Clive Beddoe that he suspected the aerospace company's accountants were fabricating time sheets on the WestJet job. "They were obviously intending to use them as a basis for fraudulent billings," Beddoe says, "and we wouldn't get our airplane out until we paid them and then we'd have to sue them to collect our money." With Beddoe's blessing, Davies, Steve Ogle, Tim Morgan, and others on the scene began a secret operation to extricate the aircraft from Santa Barbara's clutches and fly back to Calgary. Over two weeks, they surreptitiously removed sixty-five boxes of the airplane's records from the overhaul company's office and hid them in their hotel rooms. On November 28, Davies announced that he was ready to take the 737 up on a test flight. The company filed a flight plan for a round trip returning to Santa Barbara. Morgan had arrived with chief pilot Bruce Flodstedt, maintenance director Vic Lukawitski, and quality-control inspector Brian Bessem. They had secretly filed a plan with the Federal Aviation Administration (FAA) to ferry the airplane from nearby Santa Maria, California, to Calgary.

The scheme had Morgan filling the plane with fuel on the pretext that he wanted to see if the tanks leaked. He tried to discourage any Santa Barbara employees from coming on the test flight, but the company's chief mechanic and an avionics technician insisted on being aboard. Half an hour before the flight, a suspicious quality-assurance director cautioned: "It's against the law to take our people to Calgary."

"I don't have a clue what you're talking about," Steve Ogle said. He and Davies were staying behind as a diversionary tactic.

The WestJetters had planned on telling the Santa Barbara people that a light-indicator problem was forcing them to land at Santa Maria airport (and in fact the warning light came on of its own accord). When Morgan shut the engines down there, he told them the airplane was en route to Calgary and they had to disembark. The two passengers—disgruntled employees—said they'd rather fly to Canada. "Here's 200 bucks to grab yourself a cab and get back to Santa Barbara," Morgan told them. He later learned that the men blew the money on Canadian whisky at a bar, toasting WestJet all the while, and then called their masters to come pick them up once they knew the airplane was aloft again. They were promptly fired—and then rehired the next day because Santa Barbara Aerospace needed witnesses for a possible lawsuit against WestJet.

After the Canadians took off, an air-traffic control official radioed to say there'd been a strange phone call about the flight and he wondered what the heck they were doing.

"It's a very long story," Morgan assured him, "but it's all on the up and up."

"Well, we're not the police," came the reply.

But police—in the form of the Federal Bureau of Investigation—had been called in and the FBI had notified the FAA that two Americans had been kidnapped and were aboard an aircraft flying illegally to Canada. The FAA passed the concerns on to air traffic control, which decided to believe Morgan's reassurances. Later Nav Canada, which operates Canadian civil-aviation services, began asking questions of Clive Beddoe, which he managed to deflect as the 737 crossed the border and landed in Calgary.

Back in Santa Barbara, Davies and Ogle had found records that demonstrated all the discrepancies in the American company's work and took them to an unwitting clerk who allowed them to make copies, which Ogle stuffed in his shirt. Back in their hotel, the pair asked the desk clerk to tell any callers that they'd checked out. WestJet had five rooms booked; Davies and Ogle moved themselves and their cache of boxes into two rooms registered under their colleagues' names. The clerk later phoned the Canadians to say several people were inquiring about them. "She thought we'd been carousing and there were some angry husbands looking for us," Ogle says.

Renting a truck from an agency just across the road from Santa Barbara Aerospace, he and Davies spent three days driving the sixty-five boxes to Vancouver. A couple of weeks later, Santa Barbara's curious sales manager called to ask only one question: "Where were you? We went to the hotel and you'd checked out of your rooms."

"Yeah, we checked out of *our* rooms," Ogle said, and the man began laughing.

In 1999, newspapers across Canada ran headlines such as "WestJet Accused of Hijacking" over a story about Santa Barbara Aerospace's $2-million suit claiming the carrier had hijacked the 737 to avoid paying a half-million-dollar repair bill. It also accused the airline of kidnapping its employees and engaging in fraud and racketeering by threatening to take them to Canada. Beddoe admitted that his staff had ferried the aircraft to Calgary without permission, but denied they'd nabbed the crew or engaged in any other illegal acts. The FBI, investigating the claims by Santa Barbara, ruled that it was a civil not a criminal matter, and wouldn't press charges. "They tried to impound the plane to blackmail us," Beddoe said at the time. "We refused to pay, so we flew the aircraft home. It's now up to the courts to decide." WestJet was suing the American company for $10 million in return. But the case died when Santa Barbara collapsed into bankruptcy.

For Steve Ogle, the lesson learned was: "Don't screw with us; we're going to figure out how to do it right." Tim Morgan says, more mildly, "We like to drive our own destiny—and do what it takes, within reason."

6

BUILD A FAIL-SAFE
INTO THE FLIGHT PLAN
Giving employees a voice

A Canadian Parable: Air Canada and WestJet agreed to compete in a dragon-boat race. Each team practised long and hard to reach its peak performance. On the big day both rivals felt ready. Yet WestJet won by a mile. Afterwards, the Air Canada team felt disheartened by the loss and corporate morale plummeted.

Management, determined to find the reason for the crushing defeat, hired a consulting firm to investigate the problem and recommend corrective action. The consultants' finding: the WestJet team had eight people rowing and one person steering; the Air Canada team had one person rowing and eight people steering.

After a year's study and millions spent analyzing the problem, the consulting firm concluded that too many people were steering on the Air Canada team and not enough were rowing. As race day neared the following year, the Air Canada team's management structure was completely reorganized. The new model: four steering managers, three area steering managers, and a new performance-review system to provide work incentive for the person rowing the boat.

That year WestJet won by two miles. Humiliated, Air Canada laid off the rower for poor performance and gave each of the managers a bonus for discovering the problem.

—An e-mail making the rounds on Canadian computers
in spring 2003

They came after the flight attendants first, thinking they would be most vulnerable, most willing to listen to their promises of a better working life. And in some ways the unions were right. The attendants might be easy pickings. Their basic wages were low relative to their counterparts at the major airlines. Their lifestyle was fragmented; they were always on the move, dislocated from WestJet's office-bound employees. And there were some communication problems that their managers were not always sure-footed in fixing.

But what the union recruiters hadn't counted on—to start with—were the profit-sharing and share-purchase plans that increased the attendants' take-home pay along with their sense of self-worth. These incentives were combined with the empowering culture of a company that tried to please its people first and only then its customers. And, perhaps most pertinent, the airline had an executive team that recognized a real need for a forum that would allow aggrieved employees to be heard. How much better if that forum could be an arm's-length organization that was created and run by the employees themselves. An organization that would have some of the virtues of a union (support in times of trouble) without the drawbacks (empire-building and us/them confrontations). "Unions have national or international agendas—not WestJet agendas," Clive Beddoe says. "The reality is that our employees would simply be a pawn in their greater game."

Smart pilots make sure there are backup emergency measures, fail-safes for when things go wrong, to save them and their crews from disaster. In the corporate culture, one of the worst things that can happen is for employees to have no viable avenue to voice their concerns about the course their company is heading and the methods their managers— their pilots—are using to get them there. Clive Beddoe in particular was concerned about this happening at WestJet and pondered the kind of vehicle that could alleviate it. Unions weren't at the top of his list. The company attracted many people, pilots among them, who'd had embittering experiences with labour organizations during their careers.

The Canadian Auto Workers (CAW, which in 1985 had swallowed the Canadian Airline Employees' Association—CALEA), had come sniffing around within the first six months of WestJet's life. "Somebody gave the union our home address and phone numbers," says flight attendant Maria Haswell, who came home one day to a hand-delivered letter from the union in a plain manila envelope. "Everybody was totally ticked off by the conniving way the union went about it." The attendants were brimming with loyalty to their company even before the bonds deepened during the voluntary grounding; few signed up in support of a union.

The second approach came from the Canadian Union of Public Employees (CUPE) in 1998. A male flight attendant was involved in the organizing and Tim Morgan and Don Bell spotted him at a hotel where employees had been invited to meet with union officials. "We accosted him and he was very embarrassed and it seemed to die there," Bell says. The drive to sign up the attendants ended soon after. Was it a close call? "*Close* would imply 49 per cent," Beddoe says passionately. "I bet if they got 2 or 3 per cent of our employees ..."

But if not unions representing employees' interests, what then? Beddoe had talked about an employee-driven organization for two years before he could convince his predominantly anti-union staff to examine its potential merits. Clichéd as he might sound—and his beliefs were shaped by his background in bureaucracy-bound England—he wanted WestJet people to be part of the solution, not the problem, and to help manage the company from the bottom up. In other words (paraphrasing the "Canadian Parable" that made the rounds of the Internet), to have many people rowing together in the same direction while requiring only the least number possible to steer the course. "And we don't want them to feel they're a pawn of management."

While at Canadian Regional, Tim Morgan had met a contract expert working for the Canadian Air Line Pilots Association (CALPA), the trade union representing several thousand pilots across the country. He ran into Rob Winter again in 1995, while WestJet was in the making. Talking to him about employee empowerment, Morgan said, "One of these days we might need you." That day came three years later.

Winter, whose father had worked for Trans-Canada Airlines, started out in Regina slinging baggage for its successor, Air Canada. Rob moved to Vancouver in the early 1970s to play junior football and become a counter agent for the airline. He was the new guy among 125 agents with an average seniority of twenty-five years. "There was not much going on from the neck up," he says of his fellow workers. "They were miserable as sin, just dragging themselves around." He and a colleague devised a new schedule that would allow the agents to bid on their shifts by time and function. After heavy resistance, they voted to approve the innovative timetable. "That summer, Vancouver had the best customer-service turn times. And when people came into work, they actually smiled." But at a meeting of their union, CALEA, the business representative from Mississauga said the shift schedule was illegal because it didn't conform to the collective agreement.

"Why can't we write it up in a letter of understanding?" Winter wondered.

"No, the union won't allow it," he quotes the rep as saying.

"I thought *we* were the union."

Then during his job-performance review at the airline, his supervisor said Winter was being paid to work, not to think. Angry at both sides of the labour/management divide, he was elected chair of the union local and when he left Air Canada soon after, to join Pacific Western Airlines (PWA), chaired a local there too. In 1984, as the airline industry was on the verge of deregulation, he became manager of labour relations for PWA, a company he'd found friendlier and more flexible in employee relations. As it merged with several carriers to become Canadian Airlines International, Winter spent two years stick-handling overlaps in the collective agreements of different unions. He also chaired the Air Transport Association of Canada's labour-legislation committee. After leaving Canadian in 1991, he eventually wound up as a contract administrator with the Canadian Air Line Pilots Association—where he met a Calgary representative, a pilot named Tim Morgan.

MAKING A PACT

By 1996, the trade union had fallen apart and Winter became a freelance management consultant in Vancouver for unionized and non-union companies, mostly in the forest industry. In early '98, a WestJet pilot named Dave Riddell who knew him from CALPA called to say that some of the airline's flight attendants were talking about union certification. Winter spoke to the airline's Therese Harvey, the attendants' employee rep, who said: "Some of the people have been unhappy and thought maybe they should get something like CUPE. We think we're better off with an employee group with a facilitator or mediator."

Clive Beddoe had been quietly discussing the fact that as WestJet grew, so would its communication problems when

123

people no longer knew each other on a first-name basis. The most immediate challenge had been the approach of the union, which tapped into the disquiet of flight attendants concerned about opportunities to swap shifts with one another and the loss of income when their overtime pay was averaged over several weeks. Rob Winter came in as a consultant, flying into Calgary regularly from his home in Vancouver. "I very quickly flushed out the fact that the HR [human-resources] people weren't really being candid with the flight attendants. They weren't coming clean on averaging. We broke the back of that in one day. There was a lack of trust—and there was a good reason for it."

After the attendants agreed to consider using him as an in-house facilitator, Winter met with WestJet's founders. With the pilots talking about having their own organization, Beddoe said he'd prefer dealing with a single company-wide employee group.

"Do you care if it's a union or an association—or what?" Winter asked.

"Whatever the employees want," Beddoe said, hoping against hope it wouldn't be a union.

Winter introduced the idea casually to people in the Calgary headquarters. Louise Feroze, then a receptionist, remembers this white-goateed guy saying, "Hi, I'm Rob. We're thinking of putting together some kind of employee thing." More formally, he helped organize a local committee called the WestJet Internal-Stakeholders Network (WIN) with the slogan "Profit-Related Income Depends on Everyone!" (PRIDE). Part of its mandate was "good employee relations with pride in the company and its services, balanced with interesting and worthwhile work." The first meeting included the agenda items "Are we still having fun?" and "List the three worst things we do."

Winter then visited nine of the airline's stations across the country to meet employees—with none of their managers present. He brought along a draft document from the Canadian Labour Relations Board on how to start a union, just in case. In at least fifteen sessions over eight months, he asked, "What do you think would be good for you—existing unions, your own union, or ... ?" He recalls, "Not one person said that they were interested in any of the union options. A big part of WestJet is people trying to run the business a little differently, and that was very ingrained in their minds. They had ownership. They were hands-on."

Only one group met him with suspicion: the maintenance engineers in the Calgary hangar. "I showed up at night and met two guys in conversation and said I was here to talk about employee relations. If they had had a pot of tar and feathers and a rail, I would have been on it. They were really uptight. They gave me the bum's rush." Speaking with maintenance director Vic Lukawitski and mechanic Tom Woods, he learned there were issues of trust in the department—among them, charges of favouritism and lack of communication.

Winter met again with the same two mechanics who had cold-shouldered him. "I understand you've got some issues. I'm not leaving; I'd like to get to the bottom of this. Ninety per cent of the company is interested [in creating an organization], so what's it going to take to get you interested?" A mechanic named Barry Sawatski became his intermediary. "We all trust Clive," he said. "If he'd come and talk to us, maybe that's the way to go." With no hesitation, Beddoe agreed to meet them. "I'm here; let me have it," he told a group of about thirty mechanics squeezed into a boardroom. Their concerns ranged from employee selection and training to communications issues and their annoyance with the purchase of a particular 737 that demanded too much physical and financial investment to keep running. Agreeing that the company hadn't anticipated spending so much on the airplane, Beddoe said, "You guys are doing a helluva job and I'm glad you're upset. But we haven't done everything wrong. Even though the extra time and money was spent, we were still able to attract a huge amount of revenue with it. If you look at what the airplane ended up producing for us, I think we did a pretty good job." When he gave them the figures, Winter says, "it was like turning on a light switch—they had a perspective. Ttheir hard work had paid off."

In February 1999, four maintenance engineers were among the thirty-eight people who showed up for the first meeting at a hotel, safely away from the workplace, to discuss a company-wide, non-management organization. Beddoe, Tim Morgan, and Don Bell spoke to them briefly before leaving them alone to define it. Encouraged, Winter facilitated a second session with the founders and Sandy Campbell, who spoke about a recent salary survey. Beddoe outlined his vision of WestJet's future and how an independent employees' organization could play a part in it: "This is your idea, your business—you form it the way you think best. Don't ask us how to do it. This is your deal."

Somebody asked him: "Do you want us to be a watchdog for management?"

"I wouldn't frame it that way," Beddoe said. "I expect our managers to manage for the good of the company. I wouldn't focus on it, but if your group happens to uncover some weaknesses as a by-product, I think that's a good thing."

It was at that meeting, as the employees brainstormed, that reservations agent Donna Laitre—who'd once trained agents for Air Canada—came up with a name for the organization. WIN hadn't won anyone over. Laitre says the employees in the call centre, encouraged by their supportive manager, Ro Imbrogno, and their team leaders, knew "we needed to have an avenue to take employees' concerns to management. These are employees who can't talk to a team leader. We needed an alternative if someone didn't listen. We were trying very hard to come up with a different option than unions." Laitre realized that WestJet people had to take the initiative in banding together to address the major issue, communications. Why not call the embryonic group the Pro-Active Communication Team—which had the appropriate acronym PACT?

To make his task more manageable, Rob Winter asked that PACT have a Group of 7, or G-7, a septet of employees who communicate with senior executives. The G-7 reps are chosen by their peers to represent each of the company's seven large employee groups and each of them has a back-up; when they all meet together, it's called the Group of 14. The G-14 draws on associate representatives from individual grassroots departments (Res Voice, for instance, is the call centre's group).

Louise Feroze, the receptionist who'd embraced the concept from the first, admits, "There was a fair bit of un-certainty at the beginning"—some employees thought offering their opinions in public could backfire. Their facili-tator recalls, "The maintenance people were still very leery. George Knowles, their first representative, wasn't particular-ly sold but thought it better to be on the inside. He became the first chair of PACT." Winter chuckles at the memory. "And the pilots were a little leery. We hadn't built up a lot of trust at that time. I said, 'This is a work in progress. If we get it right, good. If we don't, let's change it.'" The mainte-nance reps insisted on an escape clause if PACT failed to pan out. Winter didn't want to make it too easy to pull out of the organization. Everyone finally agreed that it would require a vote of 75 per cent of a staff group to withdraw—the same percentage it took to create PACT. One industry observer has told *Canadian Business*: "In effect, [Beddoe] has locked them in. That's a very important element of WestJet's corporate structure, given Canada's propensity for strong labour movements. Here is a company that will be forever union-free." In fact, it wasn't the CEO, it was the employees themselves who established such a high threshold for change.

Now the organizers of the union-free PACT had to get the plan endorsed by all the other WestJetters. A working group prepared a question-and-answer sheet about the organization and distributed it, along with a copy of the constitution, to employees across the country. Each department called its own meeting to discuss the concept. There were critics: "I can walk into Clive's office whenever I want; why do we need this?" In what Winter saw as a show of camaraderie and goodwill, some of their colleagues took the time to explain the anticipated benefits. The reps borrowed ballot boxes from the City of Calgary and hired a notary public to oversee a secret vote in May 1999. Nearly 92 per cent of the 608 employees voted in favour of PACT, even though each was to pay $65 a year to fund its operations. WestJet would share the costs of running the organization until 2002, when the employees themselves asked it to end the subsidy to lower the costs of *their* company.

The G-7 representatives began meeting every two weeks, the G-14 once a month. In its first year, the organization asked to study the company's annual survey comparing WestJet salaries to those in the rest of the industry. "There wasn't a lot of information on how salaries were arrived at,"

Winter says. After a senior executive tried to withhold the results of the survey in spite of Beddoe's promise to release it, "We eventually got the document and found they weren't using current collective agreements and comparing apples-to-apples jobs. An entry-level airport job paid a minimum wage and their survey said this should go up to $13 or $14 an hour. We asked if anybody else had that job in Canada. Nobody did. The G-7 and myself, working with [an outside salary consultant] ended up saving the company $8 million over a three- or four-year period, based on future hiring. A lot of the jobs were way overrated. The employees said, 'We know that person who has no skill gets $8 an hour in an after-school job; why would we pay that person $13 or $14?' It was their company and they wanted to save money where they could and share equitably."

Feroze, elected to the G-7, says, "The salary consultant couldn't believe a company was working cooperatively—just opening up that process for its employees. There were a lot of meetings, at least three weeks, eight hours a day, working on the task. The pay scales were harmonized—reservations, inflight, and the airports were brought in line—so there's a relationship between the three front-line elements of the company."

She eventually became PACT's full-time administrator. The current chair, also full-time, is reservations agent June Fiori. The two women couldn't be more different: Fiori, a blonde with a broad smile, enthuses; Feroze, with lengthy black hair framing her elongated face, speaks with the precision of her past as a radio broadcaster. Fiori prefers people; Feroze likes documents and organization. Both see PACT's G-7 as an umbrella group where larger issues are addressed, preferring that the G-14s and the department groups that feed into them solve problems at the grassroots level.

Reservations agents had a pressing issue about their night-and-day work schedules, for example. The Res Voice group felt it could be handled by a task force of peers, rather than at the higher reaches of PACT. After surveying employees for feedback, the group presented the results to Deanna Coyle, the scheduling manager, who explained how jobs were timetabled and encouraged new ways of looking at the issue. A dozen agents from the call centre took on the task, explained their recommendations in detail to their colleagues and, at this writing, the revised scheduling was about to take effect. "It was agents building this system," June Fiori says.

Progressive discipline is one area where PACT will support an individual, if called upon. In WestJet's world, a person being disciplined will usually first receive a "letter of expectation" describing what the company requires of the individual's conduct. That's to be followed, if necessary, by a verbal and then a written warning. If the problem continues, a one- to three-day suspension may be imposed. (And the truth is, if a person with a persistently un-WestJet attitude is still on a three-month probation, he or she will just be let go.) PACT people like to be involved from the start as a support through any formal disciplinary process—"to make sure the company is treating the employee fairly," Feroze says. A PACT representative will take notes at an employee-manager meeting and clarify whether the employee knows what's happening and has any points to make. Unlike a union, she says, "We support them ... We're there to make sure the process is being followed."

Pilot Ben Atkins's vision of PACT was, "from the first, let's make management accountable. I don't think there's any case where any employee has been let go improperly." There was one case where a customer-service agent at a station outside Calgary was accused of theft. Although she denied any thievery, the company fired her. PACT got involved and learned that the proper documentation hadn't been done to prove the accusation. The employee was reinstated.

Rob Winter agrees that "in the early days, some people were terminated for not-just-cause and we had them reinstated." He later led PACT reps, managers, team leaders, and vice-presidents in a series of "hard-skill" training sessions. "WestJet people know how to be nice to one another; that's why they were hired," he explains. "But one thing they found hard was how to deal with problems: people not coming to work on time, even issues of theft. That's not a big part of the WestJet culture—to confront people." (One observant employee says that continues to be true: "WestJetters are afraid of confrontation, and people are squeezed out, especially those who are old-timers and are being overwhelmed by the new demands of a growing company.")

Winter would go on staff in late 2000 as director of what WestJet had labelled the People Department—human resources—and then part-time the following April until the summer, when the commuting between Vancouver and Calgary became too onerous. As a consultant once more, he continued to do annual workshops for new PACT representatives and began developing a deeper role in the organization again. He now attends every G-14 and G-7 meeting as an adviser ("You don't know what you don't know," he told the reps) and hopes to introduce such concepts as arbitration as part of the corporate disciplinary process.

Not long ago, the effectiveness of such an employee organization proved itself dramatically. Clive Beddoe approached a compensation and benefits subcommittee of PACT just as a promised pay increase was to kick in at the beginning of May 2003. Pointing out that the Canadian airline industry was in a serious state of flux and a war in Iraq was looming, he asked if employees would agree to either defer the raise for a time or suggest alternatives. The PACT reps had all the relevant numbers about operating costs and revenues. Making a huge leap of faith—"We are WestJet"—they agreed to the deferral. In early June, executive decision-makers took another look at the figures and decided things were economically sound enough to trigger the raise—and make it retroactive to May 1.

PACT is a work in progress, with employee groups waxing and waning in their interest, but Winter sees it as a singular, company-wide communications medium worth much more nurturing. "I have no reference to anything like PACT in North America, and if there has been, it hasn't survived," he says. "In particular, it's unique because the other airlines all have the same model of pilots, mechanics, flight attendants, etcetera, in separate groups. The pie is fractured. The policy of the Canadian Labour Relations Board has been to maintain the community-of-interest bargaining units. I don't think there was any political will to do anything different. For years, the Air Canada mechanics tried to get all the non-technical people out—but the board wouldn't let them do it. The board just had no incentive to change the balance of power and the unions haven't pushed for it."

A VOICE ON THE BOARD

WestJet has another distinction in the Canadian airline industry: an employee from the rank and file sits on its board of directors. Donna Laitre, who was both a call-centre trainer and the vice-chair of PACT in 2001, was the first one the G-7 group chose for the PACT position. But after attending only a couple of board meetings, she moved into middle management and had to resign both posts. She was succeeded by

Al Byl, a pilot who joined WestJet after the 1996 voluntary grounding (and is a cousin of flight-operations manager Ernie Byl and his brother, Peter Byl, a captain). He'd bought into a corporate philosophy that he believed was working in practice. In his words: "If you look after the employees, your guests are looked after, and if you look after your guests, you look after your shareholders. It's a pretty simple formula."

Deciding to play a director's role, "I knew it would be an incredible learning experience to sit on the board of a company doing a half-billion dollars' worth of revenue. You can't help gain insight and decision-making experience. It's a full, participatory seat on the board. I know numbers and I knew which end of the balance sheet is up. I have a business degree [in airline and aviation administration]." Welcomed warmly to a board of business people, some of them multimillionaires in the oil industry, he found them down-to-earth and hard-working, holding quarterly meetings that ran from early morning till five in the afternoon—with conference calls in between. He participated in one phone session while in his truck at the top of a logging-road hill, to get better cellphone reception, and in another while at a shopping centre in North Battleford, Saskatchewan, when called on to make up a quorum.

During his year on the board, Byl wore many hats, as a flight captain, a PACT representative, and a shareholder. As a director, he dealt with the intricate details of a billion-dollar loan application to buy new airplanes and was amused to hear Ron Greene ask his fellow directors: "Is anybody as confused as I am?" He felt the board members showed genuine concern, well beyond lip service, in asking him to anticipate employees' reactions to a decision: "Is it what we promised employees? Is it within people's expectations?" In his experience, the biggest challenge was keeping those hopes reasonable: "Having a package consisting largely of stock, people were always building up their own expectations. But they didn't see the whole picture in the industry—yields might be down because of our competitors. You can fly a full load and not make money. If we provide our people with good information, they'll come to their own conclusion."

In Ron Greene's view, "As Al Byl began to get comfortable, he found it very informative and it opened his eyes—and therefore other employees' eyes about how much effort the board puts in to make sure the business decisions are made correctly. Everybody has significance on the board and nobody is there to rubber-stamp. Later he was able to communicate things from an employee's perspective that we wouldn't even think of. For example, you know the board's dealing with the income statements and how that's going to influence the profit plan, but what happens when you have a weaker quarter—how is that going to be received? You can draw on the PACT representative to give you some insight. God, all these guys are seeing how their peers at Air Canada are being laid off. We [the directors] might be paranoid about one thing and he might be about another. If nothing more is gained than the PACT leader going back and saying 'This board is doing their best to look after your interests as well as the company's overall interests.'"

Byl's successor was maintenance engineer James Homeniuk, who still sees WestJet as "a people's company." Joining the board, he was intimidated and excited at first but in his brief time as a director has never felt his presence is merely symbolic—"if you want to sit as a token, I can think of better things to do." However, at least one senior executive admits, off the record: "It sends a good message: people say, 'Oh, we have a voice on the board.' ... It's *optics*." To which Beddoe replies: "It's not optics. The employee who's nominated by PACT is there to give the board some insights into grassroots sentiment and to guide us on how things will be perceived." Homeniuk views himself as a fully involved director. "I provide the people's role to the board, the impression of culture, feedback, possible anxiety in the challenging turmoil of the industry. The challenge right now is how WestJet's business model is going to outpace the rest of the airline industry."

FAITH IN THEMSELVES

The flight attendants would face a renewed test of their fidelity to WestJet five years after the previous union approach. In mid-2003, there had been communications problems between them and their inflight supervisors. One major issue was "deadheading," the periods when attendants returned to the Calgary base after their assigned flights, still in uniform but not serving passengers. A revised payment system replaced "duty time"—all the hours they spend away from the base—with "block time," which didn't include their deadheading and airport time but did pay double for the hours they actually worked during a flight. Attendants weren't yet convinced this was an improvement. In fact, the system proved so difficult to explain that a supervisor who tried to demonstrate to an interviewer how it would benefit employees used an example that inadvertently disproved her case. Embarrassed, she vowed to send out a better explanation of duty- and block-time pay rates to the flight attendants. And Don Bell agreed that the topic was now probably a good one for the regular Fireside Chat exchanges he holds with different groups of employees.

When the labour movement came calling for the third time, it was a Calgary local of the Teamsters Union approaching attendants with recruitment letters mailed to their homes.

135

The letters began, "I am sure you are aware that the airline industry in North America has been hurt by world events in the last two years starting with the attacks on 9-11 and most recently with the SARS out-break, making it difficult to be profitable. Even in these adverse conditions Southwest Airlines reported on June 6, 2003 that it was profitable with its work-force being 81% unionized."

(In fact, not long after, Southwest flight attendants began demonstrating at airports after fifteen months of negotiations with the airline had failed to settle a dispute over wages.)

The Teamsters letter went on to say: "Our members join together with a common goal—To Build A Better Future For Ourselves And Our Families. We have been approached by WestJet employees who are interested in organizing into the Teamsters Union ...

"Over the past months WestJet has told you that you can expect close to 20% of your base pay to be paid out to you in Profit Sharing. Unfortunately due to increasing fuel prices, the end of fuel hedging, increased competition and world events the great profits shares of the early WestJet days may not be achieved."

(Little more than a month later, WestJet announced its second-largest quarterly profit ever: $14.7 million.)

Then the letter took on the airline's employee organization, PACT:

> ... Progressive Discipline at WestJet is a system with no set standards or measures. Warnings and suspensions are given without reason. There is no process in place to determine what level of discipline will be used, often times resulting in different measures being taken for the same infraction between various employees. Do you know how many lates will result in a written warning or a suspension that will affect

your pay and career?

You have been told that PACT is in place to represent your interests:

PACT has NO AUTHORITY in disciplinary meetings

PACT has NO AUTHORITY to negotiate for better wages

PACT has NO AUTHORITY to determine policies and procedures

(The airline has a twenty-four-page set of guidelines on job-performance issues that points out how team leaders should proceed sensitively through the steps of a progressive disciplinary process. But in the WestJet way, it stresses that "one of the best tools you can use is to provide ongoing positive reinforcement ... Progressive discipline is only a very small part of performance management and as leaders you will find that more time spent in proactive positive leading will provide much better results from your team.")

The union's recruitment letter went on to reassure the attendants that union membership wouldn't mean they were working against their company (although unions recently had protested vehemently when Air Canada president Robert Milton and some of his managers did symbolic shifts at the Toronto airport to alleviate passengers' anger over long waits due partly to layoffs of ticket agents). There was a final reassurance: "All requests for information are held in strict confidence. WestJet will not know that you would like to learn more. We will never contact you at work!"

The letter inflamed one WestJet flight attendant, who wrote a heated reply, parroting some of the Teamsters' language:

I am sure that you are aware that every major Airline in Canada has had numerous problems to being successful and some do not even exist here today due to the "UNSUCCESSFUL NEGOTIATIONS" imposed on them by Unions.

WestJet is NOT made up of men and women like you. That is why for the past 7 years WestJet has made a profit each quarter, ranked # 3 in the World for Customer Service in the Airline Industry, won numerous awards, and so on and so on. WestJet is a family environment. We have a very exciting and rewarding future for ourselves and our families. I believe that without our employer behind us they would not be successful and without us behind our employer we would not be successful. We DO NOT need to be destroyed by your "NEGOTIATING", "BARGAINING" and "COLLECTIVE AGREEMENTS." It is clear in your letter, you have no idea of WestJets' beliefs or culture, if you did, you wouldn't be a Union Member. I do not want to be in an environment that has the "Workers" negotiating with the "Employer." The Union has destroyed thousands of lives and families in the Airline Industry, with them no longer being employed. I believe you are running low on "Members" that have been lining your pockets. The Teamsters Local Union No- 362 is only in this for their own interest.

Teamsters has NO AUTHORITY to have my name
Teamsters has NO AUTHORITY to have my address
Teamsters has NO AUTHORITY to invade my privacy
Teamsters has NO AUTHORITY to solicit me at home or work
Teamsters has NO AUTHORITY at WestJet

I HAVE Representation with Integrity, Honesty and Trust through PACT and most importantly, Clive Beddoe, Tim Morgan, Don Bell, and Mark Hill, our forefathers, and so on down the line. This is not something I will ever get from you. You have received my name and address through deception. How am I to TRUST anyone who would sneak behind the scenes to receive any PRIVATE INFORMATION. You say you are not against the company, then why would you make a statement in your letter that "everything is held in confidence" and "WestJet will not know we contact you", or "you will never be contacted at work".

I am an owner of this company as a shareholder and I AM WestJet. You have invaded my privacy and insulted me. You are against ME. You are deceitful, sneaky, unfair and very much UNWANTED.

In an effort to preserve MY WestJet, I demand that you remove my name from your "ILLEGAL" mailing list. DO NOT contact me in the near future or EVER.

An article in the *Calgary Herald* that August had the Teamsters' local president acknowledging that it wouldn't be easy to get a foot in the door at WestJet. "The people we talk to are either dead for or dead against ... The company is definitely bigger [than when CUPE tried to organize it in 1998] and we hear some horror stories once in a while about some employees' treatment." Yet even this diehard union official was forced to admit, "Then again, we hear good stories sometimes about employee treatment."

7

QUESTION YOUR PILOT
Managing executive mistakes

Some executives entertain lavishly on the company dime. If Clive Beddoe tried to, he'd probably get lynched. Consider what happened when the president and CEO of WestJet Airlines Ltd. threw a catered barbecue one weekend for the company's senior and middle managers at his private fishing lodge on the Bow River, downstream from Calgary. News of the party spread quickly through the company, and when Beddoe returned to work the following Monday, a WestJet maintenance worker stormed into his office. Pounding on Beddoe's desk, the employee demanded to know why the boss was blowing company profits on hamburgers and beer for the folks at head office. WestJet has a generous profit-sharing plan, and the maintenance guy figured his cut was being squandered on a soiree he wasn't even invited to. Beddoe told him not to worry. "I pointed out that I paid for the party out of my own pocket," says Beddoe, grinning. "He was a little humbled, but I congratulated him on his attitude. He's like a watchdog, and he hates inequities. That's the spirit of WestJet."
—*Canadian Business*, December 2000

Rob Bowerman, one of WestJet's original maintenance engineers, was the man who bearded Clive Beddoe in his den about his supposedly profligate ways. He was among those employees who might have wondered at first why they even needed an organization like PACT when they could walk right into the boss's office and demand answers.

Bowerman started taking flying lessons at age thirteen and had his licence three years later. A car crash at twenty made him shift his career goal from being a pilot to becoming an aircraft mechanic. He remained passionate about airplanes and was equally intense about his company. When he heard about the managers' barbecue, he thought, *We're scrambling to make ends meet and we're going to have this great big party? No, we're not that kind of company. How dare these guys go spend a bunch of money on something like that instead of putting it back into WestJet?* He adds now, "And I was a little bit choked that they didn't include everybody when everybody was working hard; it wasn't just management."

Striding into the executive suite, which in the WestJet way was labelled "Big Shots," Bowerman told a man in the anteroom, "I need to see Clive." Informed that Beddoe was busy at the moment, the mechanic said, "Don't worry about it. I'll just go in and see him."

Confronting his CEO, he asked, "Why are you doing this to us? There's no 'I' in 'team.'" In his mind, it was the employees' company and the founders had made it clear from the beginning that there'd be no "them" and "us." During the early years, "it was like a huge family. It was an infectious disease; it consumed everybody. The desire to make it work was incredibly strong." He didn't want to see that culture eroded.

"Clive said he was happy we were taking care of the bottom line. And he didn't realize that people were watching as close as they were." Beddoe explained that the barbecue had been his personal treat to reward some managers. Bowerman knew that this was a company chief who refused to take any salary for several years. Beddoe himself had paid for the antique desk and chairs in his office and WestJet had spent only $1,500 on the used $15,000 mahogany-and-chrome table in the boardroom. He wasn't one to waste money.

"I betcha we talked for an hour," Rob Bowerman says. "I was rather humbled at the end of it all. But I don't regret doing it."

If good pilots provide fail-safes for when things go wrong, good flight crews keep an eye on the folks in the cockpit to make sure they don't mess up. Crews should, of course, be loyal to their captains—but only up to a point. They shouldn't be blindly, forelock-tugging faithful in any circumstance. If pilots do something that will endanger the safety of an aircraft or upset the harmony in a cabin, they should be, must be, questioned.

As the seasoned pilot Stu McLean, WestJet's manager of technical flight operations, puts it, "This is the basic premise of any good CRM—Crew Resource Management—program (and you can substitute "Cockpit" or "Company" for the C)." He mentions the 1989 crash of an Air Ontario Fokker F-28 just after takeoff at Dryden, Ontario, killing twenty-four people. Although a surviving flight attendant had noticed the wings were iced up, she hadn't mentioned it to the pilot. "In fact," McLean notes, "Air Ontario apparently had a written policy that flight attendants were not to question the captain's authority on technical matters, which led to her reluctance to bring the icing concern to his attention ...

"At WestJet we teach every crew member that they have the right to speak up when they're not comfortable with an action or decision made by any other crew member. The skills required to accomplish this successfully are also taught. Of particular concern is the relationship of a pilot-in-command, the 'final authority', with the rest of the crew, but it does work both ways. For instance, sometimes a captain's bad decisions are driven by an overbearing flight officer or even a particularly aggressive flight attendant. The important thing is to break one of the links in the chain to disaster."

For that policy to be effective in the corporate world, a company has to encourage employees' critical responses not only to their immediate supervisors but also to the top-most level of the organization. And, in those rarefied reaches, wise executives keep a weather eye on one another and have a board of directors that's attuned to any turbulence created by negative anomalies in leadership styles and actions.

The WestJet culture—as professed and most often, although not always, practised—doesn't demand blind, unquestioning loyalty. It tries to create an atmosphere where people can question authority. Stu McLean says, "We want everyone to recognize that a chain leading to disaster is building, or has been built, and know how to break it. We expect everyone to voice their concerns. We teach the decision-maker to consider the concerns expressed to them and re-evaluate their position. If the decision or action appears to be a poor one, then it should be rectified. If not, then a reasoned explanation should be offered to the concerned crew member and an attempt be made to bring them on side. Unfortunately, consensus is not always possible." PACT is one way of trying to make sure this happens. An alternative is a direct appeal to a vice-president or president, which may be rare but isn't unheard of. Rob Bowerman's visit to Clive Beddoe is a case in point.

WestJet had only to look within its own industry for the effects of inviting employee access to the inner sanctum of a corporation. For more than forty years, Robert F. Six ran Continental Air Lines, a major American airline, as his personal fiefdom. He discouraged communication, much less criticism, from flowing upward in the company. A typical statement: "'My door is always open—bring me your problems.' This is guaranteed to turn on every whiner, lackey, and neurotic on the property." By the time he retired in 1982, Continental was in deep financial trouble, which led to bankruptcy proceedings the following year and again in 1990, when the even harsher Frank Lorenzo was running it. When Gordon Bethune became president four years later, he nursed the company back to rosy health by actively collaborating with his employees—even inviting everyone on staff to open houses in his corporate offices—as well as instituting a generous profit-sharing plan. Continental, which was once the worst among airlines in the United States for such measurements as on-time performance, was soon consistently ranking at or near the top in customer-satisfaction surveys.

Don Bell's executive assistant, Shirley Hall, has a long quotation pinned up near her desk that sums up the strength of not merely allowing feedback from employees, but strongly urging them to express their concerns to their bosses. It was written by Barbara Glanz, a Florida-based professional speaker, author, and consultant to companies such as Boeing, Southwest Airlines, and Bank of Montreal. She's the author of *Handle with CARE: Motivating and Retaining Employees*, based on research with more than 1,200 non-management employees. The quotation struck a chord with Hall:

> Whenever I do focus groups in an organization, I ask them what they really love about working there and then what are the barriers to that being the very best place in the world to work. This allows them the opportunity to get negative feelings out in the open, and it provides excellent feedback to the organization.

I then focus on their ideas for creative solutions for each of the barriers as a way of "moving on" and changing the negative energy into positive action. When employees have an opportunity to set aside or to let go of the barriers they perceive as hindering their doing their best work, it then allows them to be able to take individual responsibility to change what is within their control and to let go of the "victim" mentality ...[1]

With mechanisms in place to allow people to vent their feelings, with a culture that emphasizes personal responsibility, and with the doors to even the Big Shots' offices usually wide open, WestJet now appears to have relatively few employees who feel or act like victims.

A PALACE UPRISING

That wasn't always true in the beginning, when the financial projections were less optimistic than today's. WestJet was just beyond breaking even and it was feeling the effects of an industry in turmoil. There are reports from that time that would surprise current employees, including descriptions of a stressed Clive Beddoe standing in the call centre and yelling at the reservations agents to stop talking and get on the phones ("Never in my life have I yelled at a reservation agent," he retorts. "At a manager, at an executive, yes, certainly"). And he went through several executive assistants before settling on the highly efficient and personable Lin Walker. He applied even more pressure then to his second-in-command and other senior executives.

Beddoe was new to the experience of running such a complex, employee-laden enterprise. Like co-founders Don Bell and Tim Morgan, he had his own entrepreneurial business outside the airline and was reluctant at first to immerse himself totally in WestJet as a chief executive officer. That meant he had to have a chief operating officer. Glenn Pickard had soon left to run Bahamas Air, passing on the control wheel to Don Clark, who became chief financial officer as well as COO. Like many of the originals, Clark was hired for his skills and experience—controller at Time Air and chief accounting officer at Canadian Regional—and not necessarily for his attitude. Short, balding, bespectacled, he might have looked like a stereotypical accountant, but didn't always think like one. His colleagues remember him as having a good sense of humour and an ability to think out of the box. "He could do things on the back of an envelope—and then push the envelope in terms of his accounting team," says the current CFO, Sandy Campbell, mixing his metaphors. "I learned a lot from him."

While bringing vital knowledge of the airline industry in the west, he had a mixed reputation in the company. The call centre's Barry Tawse, who'd worked with him at Time, says the two of them got on "swimmingly." He recalls post-work drinks in Clark's office, in what came to be called Don's Bar: "We'd load up his little bar and have a chat as he debriefed us." Bell, on the other hand, says, "I don't have any fond recollections of that guy. We had a few guys who came from the old airlines. We needed to hire people with experience, people who knew something about the business. I think we quickly realized that they weren't able to live out the philosophies that we built our business on."

Mark Hill went from being a fan to a critic. "I liked him. He's a nice guy—hard not to like at first get-go ... [But] I saw a change of character in Don from a guy who was pretty straightlaced. And Clive couldn't get straight answers out of him." Sandy Campbell says, "Don was not focused on the raw meat-and-potatoes of the business" and gives the specific example of Clark's visits to the United States while aircraft that WestJet was purchasing underwent major inspections called C-checks. "He started to be gone AWOL on these C-checks and wasn't confiding to Clive how much they were costing."

It didn't help that a stressed Beddoe, impatient for better results in that first year of operation, was pushing Clark and his senior staff hard. At one point, he told his COO: "Either these behaviours improve—or you're gone." Unhappiness was spreading like contrails in the executive offices. "Don manipulated the entire executive team into believing that I was micro-managing the company and that I should go and he should stay." In January 1997, Clark tried to orchestrate a corporate rebellion to oust Beddoe as CEO.

Ron Greene found himself caught in the middle of it: "Don Clark had come to me saying he and all the other vice-presidents had issues with Clive and his management style and wanted me to know about it; all the senior guys were concerned about it. They wanted to meet with me—probably as a director who lived in Calgary and as a sizable shareholder—to air their grievances. Meanwhile, he'd obviously had people all riled up. He had me come to a luncheon with the senior management group and listen to everybody else. And he didn't show up. After I heard two or three guys bitching and complaining, I said, 'Clive's a straight-up guy and you owe it to him to meet with him in his office and tell him what you're telling me.'

"We went through that process. We met with Clive and Don Clark and I met with Clive, and of course Clive was blown off his feet—bringing a director in instead of coming to him directly. It was brewing over the weekend and we were trying to decide how to deal with the palace uprising— is everybody really this unhappy? Then Clive got a phone call from a man who'd overheard Don saying how he was organizing a palace coup, how he'd have Clive step aside and he'd be the next CEO of WestJet. He described Don to a T." Although the caller wouldn't give his name, he agreed to leave his phone number so a board member—Greene— could call and have the story confirmed. Again, the man insisted on anonymity, but said he'd been in a similar situation and wanted to warn WestJet's CEO. "This was a gift from heaven," Greene says. "It answered who instigated the palace coup"—and why the dissent had erupted.

(In the spring of 2003, a man installing a theatre system in Greene's home learned he was a WestJet director and said: "I've got to tell you: I was in a restaurant one day ..." and told him about calling Beddoe to report Clark's plan to dethrone him. "And you got a call from one of the directors," Greene responded. "How did you know that?" the man asked. "That was me," Greene said.)

Confronted after the call, Clark denied the accusation that he wanted the CEO's job. In spite of the incident, Beddoe decided to let the issue die and leave the COO in place. Why? "Because we all make mistakes and sometimes the most valuable person in an organization is someone who gets a second chance. He's either going to prove himself or hang himself." But by May, Clark was asked to leave. "He didn't remember the Golden Rule: he who has the gold, rules," says Sandy Campbell, who replaced him as CFO. Not long after Don Clark's departure, WestJet's fortunes began climbing. He had to leave his performance shares behind; today, some insiders have calculated, they would be worth many millions of dollars.

As profits soared, Beddoe would become a much mellower boss ("It was more than a transition; he's a different person than he was," Don Bell says) and would remain surprisingly forgiving of the man who once tried to unseat him: "In many situations, it takes a lot of patience to change ... But we all make mistakes. I hate to think I haven't."

Clive Beddoe was about to become complicit in a mistake made collectively by the directors and many of the executives of WestJet. It began after Don Clark's leave-taking, when Beddoe began to oversee the company day-to-day once again—but only reluctantly.

Although he asked the board to start seeking a president and CEO to replace him, no one would fill the position for nearly two years. "Clive had made it very clear to the board in the beginning that his strengths were in organizing the business, getting it up and running, but not in running a business of any significant size with a large number of people and moving parts," Ron Greene says. "So we started a lengthy process with a lot of interviews."

Executive headhunters identified several hopefuls from within and outside the industry. In the end, the directors felt the best candidate was a veteran of the air wars: Steve Smith, president of the Air Canada regional affiliate Air Ontario. They and the founders believed that airline experience was necessary—"but we were also afraid because of the original premise that most airlines were run by people with airline experience and not *business* experience," Greene says.

At forty-five, the Vancouver-born Steve Smith had a bachelor's degree in mathematics from the University of Waterloo and, unlike any of WestJet's founders, an MBA, from McMaster University. He had started with Air Canada in 1979 as a financial analyst in the maintenance department and held four different jobs in Winnipeg over two years, including marketing and sales. Returning to the Montreal headquarters, Smith was part of the design team for Aeroplan, the frequent-flyer program, and negotiated contracts with Air Canada's regional feeder carriers. Smith left the airline in 1988 to become president of Commuter Express in Toronto, which changed its name to Air Toronto, tripled its fleet to nine aircraft, and began serving ten US cities out of Toronto. When he joined Air Ontario in 1994, it had $150 million in revenues; as he left a half-decade later, the airline was expecting to generate $250 million. During his tenure, he weathered an eleven-week strike of pilots and a slightly longer one by flight attendants (who claimed they had to work fifteen-hour days with up to nine landings a shift).

Like Don Clark, Smith was highly presentable, with boyish good looks and a big, tooth-flashing grin beneath a long thatch of hair parted in the middle. Beddoe had met him during a conference of the Air Transport Association of Canada, the commercial-aviation organization in which Smith was actively involved: "He smiled very easily—very friendly, very outgoing, very good with people. Steve could be quite charming, and he certainly understood the fundamental business ... He talked about the opportunity to start with a clean slate without all the garbage that goes on at Air Canada, which he claimed he absolutely despised and was very contemptuous of that kind of management."

Steve Smith came aboard WestJet in March 1999, and Beddoe moved up to the role of chair as the airline was flying high into its fourth year with substantial revenues. Although still a private company, it was releasing its financial statements in anticipation of going public, and the results for the third quarter of 1998 showed a profit of $3 million, up from $2.8 million in the same period of 1997. Having launched with 220 employees and three planes serving five destinations, it had expanded to 800 people, ten aircraft, and eleven cities.

The airline pioneered the use of a small airport near the city of Abbotsford in the Fraser Valley, about a forty-minute drive east of Vancouver. Transport Canada had transferred the airport to Abbotsford's jurisdiction in 1997 and WestJet began the first commercially scheduled service out of there in June. "We used to operate out of an old converted garage, the old maintenance garage from the airport," Tim Morgan says. "In the middle of the holding bay, there were drains for the trucks, so we put plywood over the top of them, so when you walked into the terminal, you're walking over top of these drains—clunk, clunk, clunk. And the baggage belts were outside." Despite the primitive facilities at first, the move made eminent sense. About 700,000 people lived closer to Abbotsford Airport than to Vancouver International; it offered free parking; and its landing fees were a fraction of those at the big-city facility. Canada 3000 and Air Canada were to follow WestJet's lead, which would make Abbotsford the fourth-busiest field in British Columbia.

A *Calgary Herald* article said, "Smith defines his mandate at WestJet as maintaining the company's corporate culture and growth and making it attractive to investors when WestJet goes public at a yet-to-be-determined date. He also plans to continue the focus on 'employees, customers and shareholders in that order ... People will fly us first because of price, but come back because of customer service.'"

Tim Morgan, who didn't know Smith, had done some checking around before the appointment. In one case, the sister of a WestJet pilot, a customer-service agent who'd worked for Smith, said it would be a big mistake to hire him—without specifying exactly why. When Morgan finally met him, he says, "I got this weird feeling—it happens when you meet somebody sometimes ... He and I never really got along. I think what drove him nuts was the fact that I sat on the board of directors. So he was my boss in this office and then when I'm in a board meeting, I'm his boss. And he could never get his head around that." Beddoe insisted that the two men try to resolve their differences. "So I go speak to him and I think we have it resolved. I said, 'We're gonna get along together. I mean, we're digging a hole here, the rope's getting shorter, and I'm gonna help you out of this hole.'"

In spring 1999, Morgan had to hire a new director of flight operations. He chose chief pilot Bruce Flodstedt, the veteran of Time Air and Canadian Regional. He quotes Smith as saying, "Oh, you can hire whoever you like, just don't hire him. I don't think he's the right guy."

"I got to run the business, I need to hire somebody," Morgan responded. "You know what, Steve? I'm gonna hire him." And he did.

Smith, meanwhile, had hired Foster Williams from the petroleum industry as the company's first vice-president of WestJet's People Department. Don Bell says, "Foster was very bright, very knowledgeable, good technician, talked a good talk, but culturally just a disaster." Siobhan Vinish, speaking in the summer of 2003, recalled, "Foster was more big-company. He was a bit more divide-and-conquer versus bringing people together ... We need more process, we need more policy, we need more things than we did 7½ years ago, but we don't need 'human resources.' That's not the kind of company we are. We need a People Department, we need a place where people can talk to each other."

Early on, marketing vice-president Bill Lamberton offered Smith advice based on his own experience with a small, family-owned airline, a feeder carrier, and Canadian

Airlines International. "Steve worked for Air Toronto, Air Ontario, and Air Canada. So I'd often say to Steve, 'You and I have got a lot of similarities in backgrounds ... But I never learned as much as I did here in the last three years. Never stop learning, Steve—you've got all the good things that we need to grow. But this is totally different. This is taking all the things that we've learned, but at the same time listening and learning here ...'

"I guess you could say that I got along better with Steve than Tim did. But Steve had a background that was more revenue management, more of things that I did. At the same time, he wanted maybe to get in my kitchen a little bit because he had that expertise, right? So as a leader he would say 'I can help you,' and I'd say 'Well, we're doing okay.' But I did learn from him too." As WestJet moved from its entrepreneurial phase to more of a traditional business model, Lamberton says Smith was involved in figuring out "how we could get our structure to work as an executive team, hand in hand, a little better. Our executive team was very much a founding group and he came from outside— tough job, very tough job ... He came in with a sense that maybe we could grow beyond the entrepreneurial stage to the more structured stage. So he added some value in that sense."

As Ruthe-Anne Conyngham, chief operating officer of Air Ontario, would say later, "One of the things Steve is very big on is process, because without process, which is the foundation of a business, it's very difficult to grow a business. If you want to go the distance, you need the discipline."

Lamberton points out, however, that perhaps Smith was too ingrained in the Air Canada way of doing things. He was talking the WestJet style of talk and was "an excellent PR man—very charismatic in that sense—but couldn't follow through with the troops. [There were] a lot of misunder-standings."

Despite Smith's avowal in the *Calgary Herald* that em-ployees came first, he soon ran into problems with their organization, PACT. Clive Beddoe would reminisce later to a magazine journalist that "Steve got off on the wrong footing with PACT. He treated PACT like a union, and they resented that immediately." Louise Feroze was still a representative of the group, not yet the full-time administrator, when she had lunch with him one day at a management conference. "He and I were having a conversation about PACT and he clarified very, very firmly that it did not make decisions on things, it made recommendations. I thought, *We are truly in two different camps.* It sounded very much like 'Here's where the line is. We're management and you're employees.' Steve Smith is effective in executing his job if his job is lead-ing a company into an IPO [intial public offering]. But it's clear, certainly in retrospect, he has a very firm approach to employee relations. I remember when we were talking about an issue, he was surprised at that level of trust the compa-ny was operating on with the employees."

She and others tell the story of a meeting he had with one of PACT's Groups of 7. It happened after WestJet went public in July 1999. A pilot was asking Smith some pointed questions about company share options the president had. "Steve was a little surprised by the comments," Feroze says. "He didn't react a whole lot at the time. Then after that meeting it became apparent he felt insulted. Steve Smith made it known in a note to that rep. He was a long-time ex-perienced CEO of Air Ontario dealing with unions. We have a much higher level of trust and commitments [at WestJet]."

As Tim Morgan recounts the incident, a certain number of the new shares were assigned to a president's list and to employees. A senior accountant approached him to say that Smith had issued some of that list's shares to his family, contrary to their intent. Beddoe, informed of this, said there had to be a mistake. Morgan approached Smith to say, "Look, this isn't kosher," and quotes the president as saying he'd clear things up. "The pilots found out and they do not like Steve Smith, not one little bit. They were ready to hang him." But they agreed to give him the benefit of the doubt and ask him to clear it up at a general meeting of PACT. According to Morgan, Smith told the pilots: "Yes, my family got some shares, but it's not my fault, Lin Walker [then his executive assistant] mistakenly issued the shares to my family." In fact, Morgan says, Walker later told him she hadn't made any such mistake (as she later confirmed for this book).

A short time later WestJet was considering the purchase of a new generation of Boeing 737s. When Beddoe asked what progress was being made, Morgan approached Gareth Davies, who said he'd passed on the relevant information to Smith about three weeks earlier. Morgan says Beddoe then called Smith, who said Davies hadn't given him the material. In fact, Morgan insists he'd seen it on Smith's desk. Beddoe now speculates, "Steve, I think, felt that we were successful because of cheap airplanes. A lot of people felt that." But the founders and the directors thought that they had to seriously entertain replacing some of the older 737-200 aircraft with the brand-new, larger-capacity 737-700 series. And they weren't amused by their president's alleged delaying action.

Smith, as the new boy, had been much more eager to get involved in secret discussions that were already under way with Air Canada. "We can't blame it on Steve," Morgan says of the idea to form a possible alliance that would have the smaller carrier be a low-cost, short-haul affiliate and fly into eastern Canada for the first time. The WestJet brains trust had been quite willing to explore an earlier approach from Canadian Airlines, which it rejected, and then make a pitch to Air Canada. The initial courting happened on the evening of June 16, 1998, in a bar in Toronto where executives from WestJet and the two senior airlines were attending an investment conference. Clive Beddoe, chatting up Air Canada's chief financial officer, Rob Peterson, suggested a profit-sharing agreement in which WestJet would take over the Red Baron's short-haul western routes and leave the larger carrier free to focus on its lucrative long-haul business. Beddoe quotes Peterson as saying, "In your dreams!" Yet talks between the two companies did begin over the next several months. Only Bill Lamberton, who'd worked for blue-hued Canadian, demurred about getting in bed with "those red bastards." Air Canada, then under CEO Lamar Durrett, had come through a disastrous two-week pilots' strike in 1998—the first in six decades—that left its share price lower than when the airline had privatized a decade earlier. In February 1999, it had its first negative annual results since '93 and both Air Canada and WestJet announced they were talking of alliance.

By May, newspapers across the country were carrying a Canadian Press story reporting that Durrett hoped to court WestJet more ardently to forge an alliance more quickly. "We've got to do a better job on short-haul services in the West," he said. "WestJet is one of those things that gives us that opportunity. They have won their stripes and their spurs and have won a place in Canada's aviation market."[2] A deal with WestJet would allow Air Canada to redeploy its aircraft and energy to more profitable, long-haul routes. The article ended by noting that WestJet officials had not returned the news agency's phone calls requesting comment on Durrett's statement.

But they *were* returning Air Canada's calls and meeting in Montreal with Durrett and his executives—including Robert Milton, the newly appointed president. "It was a huge growth opportunity for us," says Mark Hill, who'd already had an interesting encounter with Milton about two years after WestJet's launch. A headhunter had called to say Air Canada wanted to talk to him about setting up a competing low-cost carrier. Informing Clive Beddoe of the overture, Hill decided to play it out and met Milton in his Toronto office. Air Canada was supposedly asking Hill to be in charge of route planning for the new airline, a role for which he had no experience. The two men danced around, pumping each other for information, and Hill quotes Milton as warning, "If you try to come east, we'll destroy you." That night, the WestJet co-founder, travelling at Air Canada's expense, bought drinks for everyone at his hotel bar and ran up a bill of a couple of hundred dollars. He never talked to Milton again—until 1999.

Given the corporate antagonism between the two airlines, the preliminary meetings around a proposed alliance were surprisingly cordial. Both airlines consulted with Canada's Competition Bureau, and they were quite close to reaching a memorandum of understanding. But the WestJet executives realized they'd have to change their operating style—adopt Air Canada's reservations system, among other things—and, in the founders' eyes, to change their culture. WestJet's keeper of that culture, Don Bell, says, "It sounded like a good idea if they ran their international long-haul business and we provided feeds to them ... I don't know if I ever thought it would work, to tell you the truth. I'm glad that it kinda died." The talks petered out as the Red Baron pulled back. Not until late that year did WestJet learn why, when Air Canada announced it had acquired more than 50 per cent of Canadian's shares.

The little airline from the west went back to doing what it did so well. Only Mark Hill, the self-styled 007, had truly profited from the sessions with Air Canada: "We were in a boardroom twice the size of this [WestJet] boardroom, where they would have stacks of paper spread out, and another little meeting room on the side. They would go out for lunch and I would kind of lean across the table ... I'd start looking at this stuff and take notes on it and think, *That's interesting*—all kinds of neat stuff about what their yields were on various routes." As *Lexpert* magazine reported, "In an astoundingly brief amount of time, Hill had become one of Canada's aviation industry experts ... During the talks with Air Canada, he acquired a reputation for omniscience." (At an industry panel discussion in Winnipeg not long ago, the theme was "Is it possible to make money in aviation?" Hill approached the podium, said "Yes," and sat down— then returned to explain how his airline does it.)

Now, with the aborted alliance issue behind them, WestJet's founder and directors were gung-ho again about going east on their own. That's when they ran into another hassle with their newish president. Smith was asked to come back to them with a business plan to fly to the rest of the country. Ron Greene says, "He kept delaying it and we said, 'We've asked you to do it and it wasn't done.' When push came to shove, he didn't feel comfortable because he was afraid we'd suffer Air Canada's wrath if we did. My understanding of his [thought process] was, 'We can do our own thing, but if we push into Air Canada's backyard, they're going to bloody our nose. You don't stand up to Air Canada.' Well, bullshit. We wanted to see the business plan and the economics ASAP. It was the board that pushed him into it."

Not long after, Morgan was among the corporate directors meeting to discuss the recalcitrant Smith's future. Most of the board members were at least initially reluctant to let him go. He had been with the company only eighteen months; a change of leadership after such a short time would send an unsettling signal to WestJetters as well as the outside shareholders. Morgan, a known foe of the president, was called out of the boardroom to see Foster Williams and Sandy Campbell. "They said to me, point-blank, 'If you don't support Steve Smith, he's out of here. You have to support him.' And I quite simply said, 'Well, I'm a founder of this company, I'm on the board of directors, if I do that I'll be lying to everybody. So I'm just gonna say what I got to say and what I've seen and what's happened and that's that.'"

In September 2000, WestJet finally asked Steve Smith to resign. *Calgary Herald* business columnist Charles Frank wrote that Beddoe "suggested the former Air Ontario executive's top-down management style did not fit with the small airline's folksy, collaborative way of doing things ... [Smith] even allowed that in light of ongoing—and apparently unresolvable—differences with WestJet's board of directors, it was probably best that he and the airline go their separate ways. Then again, that is what gentlemen say in what is euphemistically described as difficult circumstances." Don Bell observes, "Steve is a pretty smart guy. He worked very hard, but certain people just didn't like him. I didn't like him because he came from Air Canada." Smith took a temporary position as executive-in-residence at the University of Calgary's Faculty of Management. He would later rejoin Air Canada as president and CEO of its low-cost carrier, Zip Air, which took aim directly at the heart of WestJet.

To quell outsiders' fears about the sacking of the president, the founders had to pull together and re-commit themselves to the company. "Clive realized he had to come back permanently; he couldn't come back for a year or two," Don Bell recalls. "I remember Tim and Clive and I having a dinner and each of us committing for five years." This time, Beddoe took a salary, $200,000 a year.

There were many lessons for WestJet out of this experience. If the broad one is to question the pilot, its corollary might be to pose the right questions—do all the due diligence— well *before* the person is hired. Dig deeply enough into the background and conduct enough probing interviews to reveal the character of the candidate. WestJet has made this a priority in enlisting its front-line staff by conducting a series of interviews designed to identify positive traits such as a capacity for teamwork. Stu McLean, as both a pilot and a flight-operations manager, has identified the qualities that define good pilots—or anyone in corporate command: "They will be decisive; good communicators but equally good listeners; assertive but tactful and respectful of others; team-builders and maintainers—and *not* overwhelmingly egotistical or prideful." Yet, based on results, the founders and directors failed to consider all those standards in recruiting two of its top operating officers. In the case of Don Clark, the failure is more understandable. He was hired in the company's earliest stages, when the culture was still evolving and when Clive Beddoe, under pressure to produce, was less sensitive in his management style. Fortunately, the co-founder's skills in people-handling have grown along with WestJet.

In Steve Smith's case, the problem seemed to be that the board and Beddoe were so enamoured with Smith's background that they didn't rigorously investigate his style of management. To their credit, though, they did begin to question it as the months wore on. Director Ron Greene says,

161

"Certainly the thing that I missed—that later became very obvious to me—I mistook arrogance for self-confidence. He seemed a very self-confident person and as time went on, from a director's perspective the arrogance factor was very annoying. Communicating to the board only what he thought they needed to hear versus the whole story, good, bad, or indifferent. You don't treat directors as idiots."

Beddoe reflects, "I had high hopes for Steve and I kept trying to help him. So I was disappointed ... That's not to say Steve might not fit somewhere else, he just doesn't fit here ... The role of a CEO, to my mind, is not to just work hard at what you do, there must be a consistency in what you are and what you appear to be. That's not always easy. You never know what people truly are until the honeymoon period is over. So the lesson that comes from this is that to expect that you can parachute a CEO in successfully is highly unlikely. He just didn't fit in the culture, the value system."

There's a book on corporate-management practices that has a lot of currency within WestJet. It's the best-selling *Good to Great: Why Some Companies Make the Leap ... and Others Don't* by Jim Collins, formerly on the faculty of the Stanford University Graduate School of Business, now a private management researcher. One of his recommendations is: "When in doubt, don't hire—keep looking." Collins argues that you can't build a great company if your growth rate in revenues keeps outrunning your growth rate in good employees. "First, at the top levels of your organization, you absolutely *must* have the discipline not to hire until you find the right people. The *single* most harmful step you can take in a journey from good to great is to put the wrong people in key positions. Second, widen your definition of 'right people' to focus more on the character attributes of the person and less on specialized knowledge. People can learn skills and acquire knowledge, but they cannot learn the essential character traits that make them right for your organization."[3]

Good to Great wasn't published until 2001, but the founders and directors of WestJet knew all this stuff well before then. In a couple of crucial appointments, they just didn't apply it as well to the executive offices as they had to the front lines. In the case of Steve Smith, Clive Beddoe acknowledges that while the company did much of its due diligence—by researching Smith's reputation with his peers in the industry—"in hindsight, we didn't get down to the people [who'd worked] on the line below him. The pilots hated him, and we didn't know that."

8

FLY IN CLOSE FORMATION
True teamwork within and without

WestJet's success can be attributed to its ability to identify a unique set of core values, which the company displays on its Web site (www.westjet.com), that distinguish it from its rivals. It then focused on turning those values into action. Interestingly, Air Canada's mission statement and goals are not readily identifiable on their Web site (www.aircanada.ca).

> —Chris Bart, professor of strategy at McMaster University's DeGroote School of Business, in the *National Post*, September 2003

The ill-fitting Foster Williams, head of the People department, left soon after Steve Smith did. Rob Winter, PACT's facilitator, took his place for a few months until a permanent replacement could settle in. When one did, he was a surprising choice. His manner of arrival was also odd, with a WestJet twist. Fred Ring had been principal of Calgary's Western Canada High, which offered bilingual and international baccalaureate programs and an educational atmosphere that inspired nearly 70 per cent of the two thousand students to go on to university each year. In 1997, four senior students from the school had died in an avalanche during a ski outing in the mountains; one of them was Clive Beddoe's niece, Aimee. During the period of grief that followed, Ring got to know the extended Beddoe family well and from that contact later approached the WestJet CEO about possibly taking some kind of job with the airline after retirement. Beddoe showed him around the company one autumn afternoon and Ring fell in love with the atmosphere.

When he pointed out his lack of airline experience, Beddoe told him, "We're not in the airline business; we're in the people business ... I don't know where you would fit, but one of the things we lack around here is grey hair."

Ring was prematurely grey at fifty-four and, at six-foot-plus, a commanding presence whether as a principal or a potential member of the People Department. He met with Don Bell, learning more about the corporate culture and discussing a possible involvement in mentoring or a similar role. Running a respected, student-focused high school had given him some relevant qualifications. WestJet's employees were overwhelmingly youthful, and he'd thrived amid the pressure cooker of young people. And for a fellow with degrees in history and education, he'd been highly entrepreneurial as a principal, erasing the high school's quarter-million-dollar debt in one year through a spending freeze. He then kept the institution debt-free and rich in students' computer technology by tapping corporations and creating an alumni association and a parents' group to raise funds through casino nights and such. In November 2000, a week after retiring from Western Canada High, Ring joined the company—under cover.

The plan was for him to observe the company's culture at the ground level: apply for a job as a front-line employee and go through the regular recruitment and training process with only a few WestJetters aware of what he was really doing. Ring made the cut in a group interview and then a one-on-one session. He was assigned to two weeks' training stints as an airport customer-service agent and a reservations operator in the call centre. On his first day in class, he was the oldest person there, a good two decades beyond the company average age of thirty-four. The instructor was a former student of his who was understandably intimidated by his presence until he took her aside and told her to relax. He spent a week on the phone taking bookings, then nearly three weeks at the airport handling baggage and checking people in. The avid observer also job-shadowed specialized customer-care employees, marketing people, and maintenance engineers—for whom he worked as a gofer in coveralls on twelve-hour overnight shifts. In between, he flew to Vancouver to observe flight crews and attended open houses, barbecues, and WestJet's fifth birthday party.

His report, titled "The WestJet Way," praised the persistence of the company's culture while identifying several weak spots needing improvement. Among them were senior customer-service agents who were resistant to change and trainers who were too serious or who frustrated slower and brighter trainees by teaching to the middle. A too-common problem was communication gaps that isolated one department from another. One of his conclusions was a challenge to upper-echelon managers:

> The most important part of any training that is provided for new employees is the opportunity to meet with Clive and to hear the culture presentation from Don. This is an absolute essential that helps to perpetuate the culture of WestJet. While Clive and Don represent the heart and soul of WestJet's culture, it is paramount that all members of senior management assume a role in:

- Being visible and accessible. This will go a long way to maintaining the vision we have set as a company.
- Participating in all WestJet events. This is an onerous task but it has to be shared among all members of senior management.
- Modeling the WestJet attitude. It is important for our employees to know our senior management and to interact with them frequently.

As Don Bell says, "Fred got to learn about the company from the ground up, rather than try to learn it from the top down. He really got to see, over two or three months, whether we walked the talk, whether what he'd heard about WestJet was real or not. He is HR [human resources]. There's a lot of politics in HR, including labour law. Their role is to carry out the People philosophy in terms of hiring and training and orientation, a very important component of the business. It's the People Department that drives culture. Culture is defined by the actions of the executive; it's lived by the people, owned by the people, essentially managed by the people. To me it's always been very clear, the connection between the people and the business."

Ring started at WestJet as a consultant to refine the recruiting process before filling in as training manager. A month into his work, Rob Winter asked him, "What are we going to pay you?" Eventually Ring replaced Winter, who was filling in as the People director, and then became the department's vice-president. "A company like WestJet combines customer service and operational needs—it's almost like a taffy pull at times," he says now. "I [wanted] more focus on doing things together. The People Department was seen more as the customer-service part of the business. I saw my role as being very independent, not aligned with departments or management, almost like ombudspeople. We were a service to help everybody—management or front line. I recognized that because of the revolving door of the leadership and lack of trust, it [had been] the 'No' department ... The single biggest criterion of any leadership is the ability to develop relationships with people."

In this, he's squarely aligned with the Southwest Airlines model. A recent book, *The Southwest Airlines Way* by Jody Hoffer Gittell, a Brandeis University management professor, is subtitled *Using the Power of Relationships to Achieve High Performance*. It's what she calls "relationship competence," and the low-cost American airline has it in spades: "Southwest's relationship focus, its commitment and passion for shared goals, shared knowledge, and mutual respect, joins with frequent, timely, problem-solving communication to form a powerful force called relational coordination."[1] Among other things, its executives aim to lead with credibility and caring, invest in front-line leadership, hire and train for relational competence, use conflicts to build (not destroy) relationships, bridge the work/family divide (by encouraging employees to be themselves and create a company culture), keep jobs flexible, and build relationships with their suppliers. In other words, authentic teamwork.

ALL FOR ONE

Team WestJet, as the airline likes to call itself, tries to fly in as tight and as mutually supportive a formation as the stunt-flying Canadian Snowbirds. Mike Pritchard, station manager at the Calgary airport, recalls one flight around Christmastime when fifteen wheelchairs were required to help disabled people board their aircraft and all the agents pitched in ("The guests loved it; it was like a wagon train").

When sirens started sounding during a fierce lightning storm at the Edmonton airport, ground handlers under contract to the airline scattered in fear and WestJet customer-service agents (CSAs), flight attendants, and pilots ganged up to load the three airplanes on the ground.

Team thinking infects front-line reservations agents as well. Jeff Wimmer, the former senior manager of their department—the Sales Super Centre—has said an incentive program that rewards them with extra pay per hour was designed by a group of agents. "Out on the floor, people knew they were being represented by their own and the reps themselves became the preachers of the good news." Employees also critiqued several versions of a document to define call-centre quality in areas such as etiquette, optimum call lengths, and meeting the needs of both guests and the company. Rather than having supervisors monitor customer calls for quality, peers review and rate them, knowing that their own calls will be monitored too. Wimmer said that involving people on the firing line is a more enlightened way of doing things than at Telus, where he'd worked as an adviser on business call centres and "where decisions were made from the top down. If you wanted to change people's schedules, you did, and the message was: 'You suck it up.'"

The team-play approach is integral in hiring new WestJet pilots. Candidates have to be recommended by pilots on staff: a sponsorship form asks for details of the relationship between prospect and employee and the reasons why a candidate would be suitable ("Keep in mind the top attributes that we look for in a WestJetter: Courteous, Friendly, Outgoing, Pleasant, Cooperative, Team Player with a Can-Do Attitude"). Recruits are made to feel at home by a welcoming committee and then paired with pilot mentors. They (like other employees) who reveal personal problems can unburden themselves to their peers in an assistance group as well as to outside therapists.

Teamwork seems to pervade the pilots' realm. They planned their own voluntary overtime system, which lets flight operations know when the pilots are available to fill in for absentees. (Meanwhile, some customer-service agents do cross-training as flight attendants so they can serve passengers when crews are unexpectedly short-handed.) Those pilots trying to build up hours can take advantage of their self-designed scheme to do occasional paperwork in the flight-ops office. Pilots will also pitch in to help flight crews tidy up cabins after landing—as do all WestJetters on business or leisure travel, even up to the level of CEO Clive Beddoe—and such co-operation has saved $2.5 million a year in cleaning costs. "In the pilot group, we don't have a lot of rules about what we can and cannot do," says Tara Linke, one of 9 women among the airline's 420 pilots (and at 33 perhaps the youngest 737 captain in the world). "I go back and clean and help with the guests. It's almost standard operating procedure at WestJet. And the other day in Montreal, a maintenance fellow and two CSAs went out in the rain and pitched bags—*not* their job. I know pilots who regularly unload bags. It doesn't happen often that we say, 'It's not my job.'"

That also tends to be true in BeanLand, which Sandy Campbell runs as senior vice-president of finance and chief financial officer. "Very early on, we established Finance as a customer-service group. We had to find a way to Yes," he says. There's a bean-counting budget analyst for each department of the company. "It's good for me because if I have to know something, I can call them. It's good for the department because they've got the resource there—they know who to call and it's the same person every time. Each department owns their own budget; they build it from the ground up." Janice Paget, the enthusiastic accounting director, says in an accent still strongly defining her English background, "The departments love it. They have people who are 100-per-cent dedicated to them. People in other companies are staggered at our approach." She found England as claustrophobic and inhibiting as Clive Beddoe did and moved to Alberta as a twenty-four-year-old in 1982. Fifteen years later, after becoming a certified general accountant and running a public practice, she came to WestJet when it had only four aircraft. She embraces the airline's entrepreneurial style in financial matters: "I'm exceptionally proud of the team we have here. Whenever you see a project at WestJet, you find a bean-counter in the middle of it—or sometimes leading it. Our team is fully integrated into the day-to-day operations."

The People Department does its own team-building. One example among many is a recent outing planned by Tyson Matheson, manager of performance management, and a staff adviser, Jeff Plimmer, a former plainclothes and undercover officer with the Calgary police department. Plimmer took a group of employees whitewater rafting on the Kananaskis River and hiking up a canyon near Banff while giving them a chance to have some revelatory conversations. Other departments have used the outdoors as a bonding laboratory, from spending a day at the Calgary Zoo to whale-watching on the west coast—always with teamwork activities built in.

Company-wide, members of a team representing flight operations, maintenance, marketing, and accounting meet to make sure the annual business plan is based on reasonable assumptions. The rationale, says Tom Woods, director of line maintenance, is "to bring it from dreamland and put it through some reality checks. It's one thing to sell a product and it's another to maintain that product. Where most companies fail, they introduce a product but they don't understand the risks and exposures of implementing it."

All in all, "It's a strong team environment; nobody works alone at WestJet," says Wayne Schneider. As director of corporate training and development, he's in a strategic position to ensure that such partnering continues. Like Fred Ring, he was an Alberta-bred high-school principal. His degrees are in political science and economics and he was a quality controller in the petroleum industry before doing graduate work in education and taking some MBA courses. Schneider, with a chiselled chin and a deep gravelly voice, joined WestJet in 2002 and was soon creating tools to teach leadership fundamentals based on the airline's bedrock goal to build esprit de corps. "We can hardly keep pace with the demand for training courses. We've just had a tremendous embracing of the Leadership/Evaluation/Assessment/Development program—LEAD. Everyone is an informal leader in the company. From Day One, you're expected to solve problems, to work positively and with enthusiasm with others, and not hand off challenges to someone else."

Schneider hired a consultant, Jeff Sullivan of People Inc. of Phoenix, the former director of people development at Southwest Airlines. They interviewed WestJet's supervisors at all levels, including senior executives, to determine an in-house definition of what makes strong, effective leaders of the airline's teams.

Potential leaders can apply to LEAD from anywhere in WestJet's world. During an interview assessment in Calgary, they're presented with a scenario that might be anything from a personal conflict to a teamwork problem and given about half an hour alone to create strategies that could improve the situation. The assessors are looking for prob-lem-solving, communication, and people skills as well as an ability to facilitate the results of a solution. Two managers then meet with a candidate for an hour to identify his or her competencies—capabilities such as decision-making, in-tegrity, and compassion. At some point, they ask for details of a complex problem the prospect faced in a real workplace situation. Schneider describes some of the questions posed: "*Why* is it a complex problem? What are some of the ambi-guities? What did you do? Why was that the best course of action? What didn't work well? How did the communication go? Were there other alternatives? What were the results?" Meanwhile, the assessors observe eye contact, body lan-guage such as crossed arms, openness of communication, listening skills, and even the evenness of breathing. One woman with great potential proved to be a poor listener— "she had her agenda with very canned responses. We want to hear authenticity." (If prospects are unsuccessful in the behavioural-based interviews, the assessment team takes the time with them to review the reasons why and then of-fers a leadership-coaching plan tailored to their needs.)

Successful candidates go on to a two-day training ses-sion of twenty WestJetters to learn LEAD basics from a pair of managers and Schneider. Among other things, they come to understand typical team roles, from innovators to imple-menters. A benchmark guide is "Leadership that Gets Results," a paper from the *Harvard Business Review* by Daniel Goleman, a consultant with Hay Resources Direct in Boston. It builds on six leadership styles springing from dif-ferent kinds of emotional intelligence.

A third phase for LEAD candidates offers seven activities for the workplace, such as writing a personal mission state-ment, completing an online time-management course, and applying team roles to individuals around them. This step can take up to half a year. Schneider and the People Department created this part of the LEAD program themselves. All of this is aimed at training team-oriented leaders with the WestJet attitude. Fred Ring says, "I believe our biggest risk is not having the right leaders to perpetuate our culture."

At the same time, the airline has deepened its general training of new front-line employees. Donna Laitre, who trains advisers dealing with reservations agents, recommended the current fifteen-day training program for agents (up from ten days). The additional time allows trainers to be on the floor while trainees take live calls before returning to the classroom to hone specific selling skills. The training is lightened with presentations done creatively (in the style of TV's *Survivor*, for instance) and a couple of theme days as the recruits dress up as nerds or tacky tourists ("Some people are not used to seeing a senior manager in a Hawaiian grass skirt").

Dale Tinevez, director of the airports operations, has posted full-time trainers of customer-service agents in every large WestJet city and has about fifty team leaders available to train in the smaller destinations. He also uses videos for group showings and the Internet for such topics as the fundamentals of de-icing aircraft. The airports department has its own team-leader mentorship program to prepare people for managerial positions in the airline's various Canadian stations even before they become available. It takes candidates through exercises in communication, leadership, organization and time management, operational performance, and performance management (in learning how to conduct and evaluate employees' job-performance reviews, candidates visit other stations and shadow other team leaders).

In the customer-service area, Laitre and others have identified obvious gaps in filling team-leader positions. "Just because you're good dealing with guests, those are not necessarily the qualities to be a good manager." And that's where the People Department's LEAD program appears to be coming profitably into play.

THE POWER OF PARTNERING

The teamwork ethos is not confined within WestJet. Until Bill Lamberton took leave of the airline in November 2003, with the possibility of continuing as a consultant, he managed his own team inside the company but worked with two groups outside of it that play a crucial role in selling WestJet. As vice-president of marketing and sales, he oversaw product development, pricing, strategic planning, competitive analysis, and route-schedule design (all of which he calls the "hardball" of marketing)—as well as the brand-building of advertising and corporate communications (the "softball" side). It's an extraordinaily long list that reflects the leanness of the airline's management operation: much of this would normally come under a company's business planning department. But there was a nice synergy between the two balls he kept up in the air. One involved research to identify a new destination city and the other the creative processes of marketing it.

To help promote WestJet destinations, he used a nationwide network of travel agencies and a Calgary-based ad agency. Both provide useful examples of the airline's style of teamwork with external suppliers.

In the beginning, Lamberton decided to ally the company with travel agents, paying them commissions—unlike the short-lived rival Greyhound Air, which initially bypassed their services and had to come crawling back after realizing its mistake. Before WestJet's launch, Lamberton and Don Clark did focus groups with agents to test their reaction to its new way of doing business. One concern was its Canadian pioneering of ticketless travel, which uses only reservation confirmation numbers instead of paper tickets. Another was a reliance on its own reservations call centre rather than the costly global distribution systems (GDS) that electronically link agents and airlines, car-rental companies, hotels, tour operators, and cruise lines. The agents balked at both ideas, especially the fact that the airline wouldn't be connected to their computer systems, but went along because they received commissions from the start. Of course, their 9 per cent was much less on a $200 WestJet round trip than on the $600 Air Canada equivalent. But what with Greyhound's decision not to pay commissions, and the majors' later clawing-back of the percentage paid to the agents, WestJet began to look good. "We've always been travel-agent friendly," Lamberton has said. With good reason: the agencies continue to account for 40 per cent of the bookings.

The airline eventually decided to use expensive GDS systems to link with the travel agencies. However, it asked the agencies to share the cost by accepting only a 5-per-cent commission rather than the 9 per cent they'd continue to get by booking travel online at WestJet's user-friendly Web site. Lamberton says the agencies generally agreed this was fair. In 2003, the year it incorporated the Worldspan and Galileo systems, WestJet was named Canadian travel agents' favourite scheduled airline in awards presented by the industry publications *Canadian Travel Press* and *Travel Courier*.

As Bill Lamberton and his people were reinventing the travel-agency side of the airline business, they were also confounding the Canadian advertising industry. Lamberton's decision to use a tiny local design and production house, Creative Intelligence Agency (CIA)—instead of a major agency with a pan-western or national reach—appears startling at first. But the more it's analyzed, the more it makes sense. As Siobhan Vinish has pointed out, "Smaller agencies have a culture similar to ours—the ability to do innovative things with limited resources."

In fact, Lamberton had no particular preference when sending out requests for proposals for a launch campaign from agencies large and small. "We certainly had our low-cost frame of mind so they were going to have to be creative and competitive. I don't think we went in saying we had to have a small agency; we just picked the agency whose material we liked best, and CIA won. They nailed our business plan with the feeling that it's visiting friends and loved ones; it's going to get people out of the cars, out of the buses, off the couch. And the idea that we were competing more with Imperial Oil and the rubber-tire traffic than we were with the monsters that are Canadian and Air Canada. That we were folksy, western, friendly."

CIA's principals, Dean McKenzie and Barry Anderson, graduates of the Alberta College of Art, had apprenticed at larger ad agencies before starting their own shop in 1981. As talented designers and production specialists pioneering in computerization, they freelanced for well-known houses such as Baker Lovick. While much of their work was in the agricultural sector—which certainly made them folksy and western—they also strategized clever campaigns for several new radio stations in Calgary and Toronto.

"We had the responsiveness and flexibility of a small company and the strategic thinking and business consulting of a big company," Anderson says. A large man who leans to cowboy boots, he looks like a football player and in fact is the son of the legendary end with the 1950s Calgary Stampeders, Sugarfoot Anderson. As McKenzie points out, "Our type of business model would appeal to a start-up company that was small, mobile, flexible but smart. WestJet's whole premise was to not be Air Canada. We fit in very well." His own image—jeans and a '50s-style cowlick— also meshed with the airline's casualness. Anderson says of CIA's low profile, "We were under the radar." McKenzie adds, "Our light was significantly hidden under the bushel." Bill Lamberton acknowledges that "in their shop, we were a big customer and they paid attention to us and learned to grow with us. Whereas in a large agency, we might have been a very small customer and have to follow [its] processes."

CIA would soon move into new quarters in a downscale building and redesign the depressing space with rich earth-hued walls, shiny chipboard floors, and corrugated-fibreglass office doors—all very uptown. The place would become even hipper with the gradual addition of a pool table and a mini-gallery of paintings and figure art (including a dancing-pig sculpture) by fine Alberta artists.

CIA established the right tone with its first print ads promising "everyday low fares" and featuring quirky cartoons and slangy language: "It's our business to jet you around the West for peanuts—every day ... We keep costs down by serving peanuts instead of meals ... We've streamlined reservations, check-ins and loading, too, thus dodging zillion-dollar computer systems." Lamberton rejected one ad showing a revolver and the tag line: "To fully appreciate what business travel costs were like before WestJet, cut this out and hold it to your head." Everything he and the co-founders approved was low-budget, price-point advertising rather than slick multimedia campaigns. Like Southwest Airlines' campaigns, the ads used humour to leaven the hard sell.

The cost of the launch marketing campaign was about $1 million. The thrust then, as it is now, was to focus on newspapers, with some magazine and radio coverage but almost no television. The exception—and the only full-production TV commercial WestJet had done until summer 2003—was an image-setting, thirty-second vignette in which a man drives into a self-service gas station in a raging winter blizzard, brushes off his windshield, sneezes his hat off, and tries to fill up, only to realize the station is closed for the night. The message: "It's comforting to know there is a better route. With WestJet, low fares are here to stay." Lamberton continued to encourage such everyman cheekiness. When Air Canada began flying into the Abbotsford Airport à la WestJet, the ads punned on the city's billing as the raspberry capital of Canada by displaying a large berry with the line: "The only big red thing that belongs in Abbotsford."

As spot-on as CIA's external advertising was for Bill Lamberton, it was the brand-building of the WestJet culture they helped do within the company that excited the airline. All parties agreed that the airline's defining feature should not be low fares, which can be matched, but the high quality of its business practices and its corporate personality—which the employees themselves had to create.

In the fall of 1996, after the voluntary grounding, CIA met in two half-day sessions at a local hotel with Beddoe, Don Bell, Bill Lamberton, Siobhan Vinish, other marketing people and department directors—about a dozen people in all. They were there to respond to four questions: Where are we now? Where do we want to be? What's keeping us from getting there? And how are we going to get there? "We needed direction from them as a company, not individual departments," McKenzie says. "They'd been through a very traumatic period in their history with a lot of intense pressure, particularly on Clive." Out of these meetings, Anderson says, "there was a real commitment to making the vision/mission/values an integral part of their business planning, human-resources planning, and marketing planning ... Don Bell says a brand is a promise." As McKenzie points out, "you can spend millions of dollars on advertising, but if that flight attendant treats a customer badly—she doesn't know what that promotion is all about—all that money is gone. A smile is a logo too. That's the manifestation of the corporation. And that's what WestJet understands perhaps better than any of our other clients."

Bell has since spearheaded discussions to define the values that describe WestJet's way of doing business and that are now prominently displayed in the airline's offices:

- We are positive and passionate about everything we do.
- We will take our jobs seriously, but not ourselves.
- We embrace change and innovation.
- We are friendly and caring towards our People and our Customers and treat everyone with respect.
- We provide our People with the training and tools they need to do their job.
- We celebrate our successes.
- We personify the hardworking "can-do" attitude.
- We are honest, open, and keep our commitments.
- We are TEAM WestJet!

Bell and his people also came up with the slogan CARE— Create A Remarkable Experience. CARE, he says, "is the branding of the culture, and the philosophy of CARE is that if you take care of our people, they'll take care of the customer and the shareholder and the company. CARE is the word that describes our values. It's about empowerment and trust, treating people how they want to be treated, about exceptional customer service, about empathy and compassion. Those are the components that make our culture and CARE is just a brand name for the culture." CIA created prominently displayed posters and even employees' identity photo badges spelling out the values and vision— "WestJet will be the leading low-fare airline that: People WANT to work with, Customers WANT to fly with, and Shareholders WANT to invest with." Every business card bears the mission statement: "To enrich the lives of everyone in WestJet's world by providing safe, friendly and affordable air travel." And employees have "culture cards" they sometimes flash at one another at appropriate moments as reminders of particular corporate values.

The tightness between CIA and the airline was demonstrated in a review of ad agencies in 2000: out of seventy on a long list, three were short-listed—two of them major shops—but WestJet decided to keep using CIA for the following four years. "We're now their largest and their anchor account," Lamberton said before his recent departure. "They have grown with us, understand our culture and our business planning—understand what makes us tick, how we are as individuals, what we need, how we work as a team." Now the next agency review has been extended until 2006 and, while there is no guarantee CIA will survive it, WestJet does pride itself on remaining loyal to those who went through the thin times with the airline and still perform ably.

That thinking holds true in other areas of the company. Bill Clinton, a certified professional purchaser who worked for Time Air and Canadian Regional, moved to WestJet in 1999. At the time, the airline didn't have a true department with all the checks and balances to oversee the buying. "It was a disjointed group," the white-goateed purchasing director told me in the summer of 2003. Clinton, who left the company later that year (to be replaced by technical services director and acting purchasing director Rob Taylor), explained, "The company started so fast and people would just react to things that were needed. Individual departments were doing their own purchasing—at not necessarily the best price—because of a lack of a go-to department ... There were so many vendors and brokers and we've narrowed them to those we deem as partners. It's not grind-them-down-to-the-lowest-possible price. This comes from the top down: we're only as good as our suppliers; if we have great suppliers, we look good too. It's not the old confrontational way."

Purchasing buys everything for WestJet except actual whole aircraft and ground-handling equipment. Someone in the department is on call twenty-fours a day, seven days a week, and—often working on a laptop at home—can have a part delivered to a remote outpost in the middle of the night. An indication of the purchasing department's importance in the company is that it reports directly to the senior vice-president of operations, co-founder Tim Morgan. Similar departments in other airlines might report through engineering and maintenance executives.

Partnering vendors have earned their place at WestJet, and the department constantly monitors these relationships with in-person meetings that result in a report card evaluating vendors' services. At one recent session, a happy vendor from the major American aviation supplier Honeywell asked, "Gee, *we* can win too?" That's because purchasing decisions don't depend on price alone. While the airline can always find a single aircraft part cheaper from one supplier than another, a preferred vendor might provide value-added services such as efficient delivery to any WestJet base in the country. An example is the two firms Data Business Forms and its Calgary subsidiary, Sundog Printers, which supply all the airline's printed forms across Canada. The companies hold the bulky inventory of forms and bill only on delivery, saving WestJet about 25 per cent in costs.

Another key supplier is Volvo Aero Corp. in Seattle, which sells aircraft parts. Volvo agreed to let WestJet pay a set amount every month for a package of parts that would be supplied as needed over twelve months. This allowed the airline to defer costs in the first quarter of the year, traditionally the slowest in the industry. "And these guys will go the extra mile for you," Clinton had remarked to me. "We needed a part so bad that we chartered an airplane and John Brooks of Volvo flew with the part in his hands on the charter from Seattle to Calgary and hand-delivered it to maintenance." As a result of such dramatic displays of WestJet-type teamwork, "They get a long-term relationship and we're not going to walk across the street to someone else next time."

Such close-formation partnering is how—to paraphrase business professor Chris Bart of McMaster University—WestJet translates a unique set of core values into successful action. The corporate mission statement focuses on enriching the lives of everyone, from employees to passengers, through safe, friendly, affordable air travel. But as Bart points out, "Any organization, large or small, can lose its long-term focus when faced with such things as an influx of new employees or the turbulent and competitive market conditions we are seeing now. Moreover, the faster an organization seeks to grow or diversify, the greater the potential for it to drift from its course." In modelling itself on Southwest Airlines, WestJet has succeeded where other carriers have failed—by defining the company's mission intelligently and then trying to put it into practice through real, ongoing partnerships with its stakeholders. For, as Bart notes generally, if any stakeholder group of any corporation gets angry enough, "it will stop doing business with the company, putting its very existence in jeopardy. Just ask Air Canada, which is in the midst of learning this lesson the hard way."

9

LEARN TO LAND ON ONE ENGINE
Productivity and empowerment
in the workplace

The measurement of revenue passenger miles created per employee is considered to be a fairly representative measure of productivity that allows one to compare different airlines with different fleets. This broad measure essentially measures how many employees are required to move passengers around on the airline ... WestJet exceeds Air Canada's productivity in this regard by a wide margin—46% in 2001 to be precise.

—Economics professor David Gillen, Wilfrid Laurier University, in *The Future of Canadian Aviation: Frills, No-Frills or WalMart*, 2002

I've helped maintenance engineers change wheels on airplanes. It takes 127 strokes of the jack handle to get the main gear off the ground to change the wheel. I never could have done that with Air Canada—it would have been a basic conflict between the union and management.

—Terminal support manager Dale Mitchell, an ex-Air Canada employee, recalling his days as a ramp supervisor with WestJet

Being forced to land on one engine—on a wing and a prayer—is never a welcome event, but learning how to do it

185

is a vital skill that must be practised. Every pilot in Canada who flies multi-engined aircraft trains for single-engined landings and gets tested on them at least annually. Stu McLean, WestJet's flight-operations technical manager, reflects: "I've had to single-engine-land twice during maintenance test flights after unexpected failures. And at one point we also were shutting down each engine separately inflight to check its relight capability. There have also been several incidents during revenue service in which a single-engine landing was required either because of a 'precautionary shutdown' or because of an engine failure. The most dramatic example of how the training paid off was during a total engine failure on takeoff out of Kelowna. The crew handled the single-engine problem expertly and by the book, immediately taking the aircraft back to Kelowna for an uneventful landing. The biggest problem is keeping the press from blowing it out of proportion."

There are times, though, when operating on one engine can be highly productive. Early in WestJet's life, its pilots were using the manuals and procedures of United Airlines, which had long since decided it made sense to have jets taxi on the runway with only a single engine running. The idea was to save on fuel costs while on the ground. With Tim Morgan pushing the concept at WestJet, all his flight crew bought into it. Today, it's standard operating procedure for all the airline's 737-200s (there's little fuel economy with the 737-700s) and, according to McLean, it saves an estimated $1.5 million-plus a year.

Running effectively on the least amount of resources— that's one definition of productivity. The American management consultant Robert Half once said that a company can't increase its productivity, only its people can. Based on results, that seems to be a WestJet tenet. The carrier operates with an average of only 77 people per aircraft compared to Air Canada's 140 (a ratio of the total number of employees to the number of aircraft). That's a major reason why throughout 2003, the Red Baron of airlines was trying desperately to extricate itself from insolvency. Industry analyst Cameron Doerksen of Dlouhy Merchant believes that despite Air Canada's attempts to rein in expenses, WestJet is likely to retain a 40- to 45-per-cent cost advantage.

The company doesn't want to take that for granted, especially given the recent growth of low-fare rivals, including spinoffs of the country's major carrier. "You should always be a little nervous in this business," Tim Morgan says. "We always have to pay attention to them, whether it's Jetsgo, Zip, Tango, anything. You have to be careful. As soon as we open the door for somebody to slip in underneath us, the game's over. You notice our costs keep coming down and there's never any room underneath us. It's not so much the fare structure ... You have to be able to produce the product cheaper than your competitor ... It's the cost-per-seat-mile that makes the airline successful and will keep us around forever." This is a measure of the cost per mile of operation—including fuel, labour, overhead—balanced against the number of passengers on board. WestJet's is nine cents per mile compared to Air Canada's estimated twenty-three cents. "If Air Canada's labour worked for free, absolutely for free, they'd still be 25 per cent above our cost."

In most areas, WestJet's productivity record is impressive. For instance, the company decided to take on the task of provisioning airplanes (with snacks, liquor, and other in-flight necessities), rather than contracting it out. By using its own employees, rather than contracting a variety of outsiders, the airline made life simpler for its flight attendants and also saved an estimated $1 million in the process during the first year. This sort of high yield from productive labour usually translates into satisfied customers. Callers booking WestJet flights wait an average of only one and a half minutes on hold. And of 1,756 complaints to the federal airline ombudsman in 2002, the airline had only 20 while carrying more than 4.3 million passengers.

The astonishing improvement in the airline's on-time performance (OTP) is due in part to the creation of its own ground-handling crew in the pivotal Calgary airport. About half a dozen key flights originate there each morning and if they're delayed, the national schedule can get badly out of whack. WestJet had previously contracted the task of loading and unloading baggage at its home base to a Canadian branch of Airport Terminal Services of St. Louis, Missouri. It wasn't working. Dale Mitchell is an old aviation hand (with a wonderful handlebar moustache) who worked in various front-line and supervisory roles for Air Canada over twenty-seven years. After a general round of layoffs, he came to WestJet in 1998 as a customer-service agent at the Calgary terminal. Noticing problems with baggage handling—occasionally more than fifty bags could get left behind in the early years—he approached his supervisor about having someone keep an eye on the ramp. Mitchell was delegated as overseer and things improved.

But during the next few years, as he moved up in the ranks, finally becoming terminal support manager, the situation with the contractor worsened. The airline kept analyzing when it might make sense to have its own people on the ramp in Calgary. In August 2002, it hired Dean Pawulski to become airside manager at the terminal to run its own ground-handling team which became known as the Turn Around Crew—the TAC team. Puwalski, a recreational pilot, started in the business as an apprentice airline mechanic and became a licensed flight dispatcher and base manager for Calm Air, operating in northern Manitoba. In his new job at WestJet, he hit the ground running, building up a team of 120 ground handlers, including the company's ideal ratio of one team leader for every fourteen front-line people.

The transition proved difficult. At the time, the People Department was so overwhelmed with WestJet's growth that interviewing enough potential baggage-handlers was a problem. (In a move that some managers in other areas of the company didn't welcome, the department had decided to take on the initial screening of most recruits and at some point be involved in the interview process of all of them.) Almost as fast as ramp workers were hired, they seemed to be quitting or sometimes had to be let go after progressive-discipline meetings. Pawulski says short-handed crews were working anywhere from eight to twelve days in a row. As Mike Pritchard, Calgary's station manager, points out, "People were leaving because we were killing them." He knew how bad the situation was through Don Bell's regular give-and-take Fireside Chats with groups of employees and because Pritchard scheduled Open Mike with Mike meetings to ask staff at the airport for direct feedback. After one session, a twenty-year-old TAC member named Bryan Fenton sent him an eleven-page complaint about, among other things, part-timers' share of shifts on the schedule. It ended: "I thought everyone was equal at WestJet." Since then, the scheduling challenges are sorting out—so much so, Pawulski says, that at a recent crew meeting, Fenton had only positive things to say to his fellow workers. Between get-togethers, Pritchard tries to keep everyone communicating through "Didjaknow" forms where people can signal problems and offer kudos: "The bag belt broke for two hours ... Johnny worked his butt off and stayed for four hours."

The TAC crew certainly didn't turn around the on-time problem alone. The airports people were part of a system operations performance team—a task force of flight operations, maintenance, marketing, and the operations control centre—charged with analyzing the roadblocks to improved OTP. Knowing that flight schedules contributed to delays, the airline made key changes in timetables. Among other factors identified and revamped were inefficient boarding procedures and a tardiness in Calgary-based aircraft taking off in the morning.

In terms of productivity, the airline's OTP statistics have increased dramatically while the amount of missing or damaged luggage—the bag ratio—has decreased. Canadian carriers aren't compelled to reveal their on-time performance, which measures how often they arrive within fifteen minutes of scheduled arrival at the gate. Air Canada doesn't release such information. WestJet began publishing its figures in June 2003, when it was first among eighteen airlines (all the others were American) with a 92.3-per-cent record to Southwest's 85.4. At this writing, its ranking has gone slightly up and down while still sitting among the top two airlines.

By the end of 2003, the TAC team had grown to 136, a dozen of them hard-working women hired to add diversity in the approach to the job. Dale Mitchell describes the current crop of baggage handlers: "There's a sense of urgency out there when they attack an airplane. There are times when we have the bags offloaded before the guests get there [to pick them up at the baggage carousels]. Getting the bags on, we try to be consistently ahead of the customer-service side—there's a friendly challenge. And if we've got everybody aboard with all their bags on, we'll leave early." Dean Pawulski attributes their productivity entirely to the corporate culture of ownership and empowerment. Sounding like a commercial for the company, he says, "When you care about your teammates, the guest will benefit in the end. When you come to work as a drone, you're going to grab the bags and throw them around and not have the passion or the drive. But you're happy to be here; you know you can make a difference. I've worked for eight companies and no one's even come close."

MORE POWER TO THE PEOPLE

Alanna Deis is director of the People Department. She has a business-administration degree from the University of Regina and a travel bug that had her working for banks in England and Australia and teaching English to teachers in Japan. Settling down back in Saskatchewan, she oversaw compensation and benefits in SaskEnergy's human-resources department. When WestJet sought someone with skills in that area as well as in systems, Deis joined the company in 2000. She describes the airline then as "a puppy with big feet" displaying numerous inefficiencies that needed fixing. One solution was to ditch a document-based computer database organizing employee information (payroll, time and attendance data, insurance claims) and adopt a more efficient Integrated Resource Information System—IRIS, for short.

An efficiency that Deis herself had to learn early on was communicating in clear English. Writing a general notice to employees about vacations, she began formally, "Further to the reinterpretation ..." She says, "I got hit over the head badly by Marketing and the VPs. The words were not WestJet enough. I was used to SaskEnergy and a union environment. Words are so powerful." The airline's penchant for plain speaking has been praised in Regina's *Leader-Post*: "One of the reasons why we like WestJet—aside from the low fares—is its ability to use clear English. At its modest headquarters, the executives' offices are marked, simply, 'Big Shots.' Accounting is 'BeanLand' and human resources is the 'People Department.' Get the idea? It's clear language for a company that carries passengers and makes lots of money. It is a delightful contrast with the many organizations that routinely mangle our language."

In the lean-but-not-mean WestJet style, Deis was soon handling much more than compensation and benefits. Her realm now includes the nuts and bolts of an employee-travel desk, the sensitive area of performance management, the broad reaches of organizational development, and—perhaps most vital of all—recruitment. As in all organizations, productivity at the airline begins with the right employees coming on board. In one recent six-month period, when WestJet was opening five bases, the People Department had about a dozen recruiters reacting to 25,000 applicants and hired about 500 of them. A deluge like this calls for an efficient handling system. "We have to make sure that the system is in place so we are treating the other 24,500 with dignity, fairness, and equity—not only because it's right but because they fly our planes and they have brothers and uncles who fly them," she says. Although there's no longer time to send handwritten postcards to those rejected, each one does get a communication from the airline.

Those accepted for interviews are first briefed in groups of about twenty. Likely candidates then have three separate forty-five-minute sessions with a team including a recruiter, a team leader, and a peer trained in behavioural-based interviewing. The interviewers always raise the "WestJet fit" issue: "Tell us about a time in a previous company when you were happy or unhappy about the level of teamwork." As Deis says, "You learn a lot about them through how they answer that. It's the perception of how they saw that world. If that isn't answered properly, the candidate doesn't have a chance." In front-line positions, such as customer-service agents, attitude is all-important. But while acknowledging that "hiring for attitude, training for skills" can make sense, Deis adds, "We've been hurt by that philosophy. We don't want to hire a real happy person that we have to train for eight years to be a tax accountant."

Members of the recruiting team document their observations and meet later to reach consensus on a candidate. If there's disagreement, the individual can return for another interview. Only after checking personal references does the team introduce the prospect to the relevant department managers. "It's not a perfect process," Deis says. "Some people can fake through an interview. Like the legal system, it's the best we have. Doing it right the first time is so important." Because of the general success of the three-interview scenario for the mass of candidates, it's becoming standard even for more senior people.

The idea is to hire positive people and keep them that way. "A lot of companies do a lot of work to make their guests happy—and the customer is always right. We say guests are right 99 per cent, but our employees are right 100 per cent of the time. If we make our employees happy, they will make our guests happy. It's not the other way around. They've got much more free rein than many other front-line employees. They're completely empowered to make guests happy."

A minor yet endearing example was reported in the *Vancouver Courier* in 2003: a husband had dropped his wife and sixteen-month-old son off at the airport for a WestJet flight en route to Moose Jaw, Saskatchewan—only to get a frantic phone call from his wife during a stopover in Calgary that she'd left the boy's well-loved teddy bear behind. It was a weekend and Canada Post and courier services couldn't get the bear to the teething child for two long days. Dad called WestJet, which offered not only to send Teddy off on the next flight, but to ship him free. "A big kudos to WestJet for doing a little boy a big favour," the *Courier* said. "The only question we have now is, will Teddy be eligible for bear miles?"

As Don Bell describes WestJet's empowerment principle, with a bit of exaggeration: "We tell our people they can give away company money. Well, there's no other company in the world that will allow you to do that. You as an agent can buy a hotel room, put somebody on another flight, buy a ticket for them on another airline, buy some roses or buy them dinner—without checking with us ... I think they make better decisions than the supervisor or a manager would. I've never seen it backfire. And I've seen people give away tons of money." As Ro Imbrogno, the customer-service director, puts it: "You take charge and you beg for forgiveness afterward." But Bell argues that "because they are owners and profit-sharers, they never give away too much."

Lynette Bryant, the reservations agent who'd taken WestJet's first call from the public, later became a customer service agent (CSA) at the airport. Recently, she had to handle the case of a man whose top-of-the-line dog kennel, complete with air conditioning, came off a flight with a cracked top. "I could have said, 'WestJet won't cover that,' but he told me he'd just paid $300 for it." She asked him to leave the kennel with her and went on the Internet looking fruitlessly for ways to have the top fixed or replaced. Finally, she offered the owner of the kennel $250 and a $100 travel credit. When the man decided to just have it duct-taped together, she gave him a $250 credit, which he said was excessive. "We'd spent a couple of days trying to figure out what to do," Bryant says. "We want to keep people happy, but we don't want to throw WestJet's money around—*our* money."

Don Bell grants there are guidelines about dishing out company cash. Reservations agents in the call centre can, in specific cases, override the regular cost of flights and decide not to charge fees for cancellations or minors travelling alone. But an agent making a habit of such decisions—revealed through routine monitoring of calls—will receive more training.

There's a fine balance between saving money for customers and earning it for the company. Early on, Hobe Horton and his measurement-solutions team discovered that "if you reward [reservations agents] for taking the most calls, that's what they'll do. But taking a call and making a sale are two different things." At the start of 1997, WestJet began a productivity differential (PD) program that offered call-centre employees a bonus for completed sales. While the agents' base hourly rate is only $10.45, good ones can bump that up by $8 and some earn as much as $23 an hour. "We were able to show Clive conversion rates—chances to sell, converted into sales—were improving. For every $1 we paid out in PD, we increased gross bookings by $40."

THE JOY OF FLEX

About a tenth of the four-hundred-plus booking operators—Sales Super Agents, in WestJet's jargon—work from the comfort of home rather than the Calgary head office. Sometimes a part-time agent sets up a home office with a company-provided PC, separate phone line, and high-speed modem along with his or her own desk, chair, and quiet space. A team leader visits such agents once a month for debriefing and the call-centre managers record a weekly broadcast updating Home Res agents on industry developments. They put in anywhere from twenty-five to forty hours a week, receive full health and dental benefits, and can fly at the employees' rate of $2.50 anywhere on WestJet. Pino Mancuso, a reservations manager, says the lesson the airline has learned is "Send your strongest people home first. It is important to present the ability to work in Home Res as a reward for great work, and so the most capable agents are admitted to the program."

June Fiori, the PACT chair, is a single parent who for two and a half years wanted to be around her children as they attended school. As a home res agent, she curtained off a desk in the corner of her bedroom. Logging on shortly after 8 a.m., she'd take two fifteen-minute morning breaks and a forty-five-minute lunch. "If there was a snowstorm, I was there; I didn't have to drive to the office," she says. "Even if I was sick, I generally worked anyway." Sometimes the home agents will lengthen their shifts when call volume is high or even handle calls on their days off. Such results justify the distant-office idea, says Jeff Wimmer, the call centre's senior manager. The company also benefits from space-saving in its call centre and increased morale— "They're generally happier."

Flexible working conditions like this can heighten productivity. At WestJet, no group appears more flexible than BeanLand, which in other companies is often a staid and unmalleable accounting department. It operates from 5:30 a.m. to 11 or 12 at night. As many as a fifth of its fifty people work odd hours that make sense for them. "We run a night shift with people filing, doing data entry, routine tasks," says Sandy Campbell, the chief bean-counter. "Why wouldn't we make better utilization of the bricks and mortar? Gone are the days of an office being 8-to-5. We're a 24/7 airline."

Payroll manager Jan Kerstiens is one of the department's originals. She took leave when her daughter was born in 1998, returned four months later, and soon decided to work nights—"from 4 till 9 or 10 and some weekends too. I probably got more done in those five hours than some people would in a day. If I was forced to come back to days, I might have left." The only question she had to answer about her flextime was: "What's going to work for you, Jan?" The policy for the full-time financial people now is to work 160 hours, anytime they want, over a four-week period. She's now in from 9 to 6 and appreciates the quiet time in the office at day's end.

Flextime differs in each WestJet division; some still stick to forty hours every week, days only. The maintenance department offers an informal system for its overnight shifts. Tom Woods, director of line maintenance, says, "If it's 3 in the morning and everything is slow, we'll keep minimum crews here and send the others home—and pay them for it. Because there's many other times we ask them to stay." To make a formal schedule change in any department, 70 per cent of its staff have to approve. The information-technology group, a male-dominated area, has had only full-time employees—until one new father requested a modified parental leave. He asked to go on an hourly rate and work weekends and evenings. "I-tech is a little nervous about it," Kerstiens says, "but they're going for it."

There's a similar flexibility in redefining the traditional roles of a finance department. Corey Wells, the director of audit and advisory services, says in other companies "internal audit is viewed as a necessity, but not a value-added area." Not at WestJet. In one case, Wells and his people evaluated the accounting department's relationships with credit-card suppliers and then renegotiated its card contracts to save about $250,000 a year. Rather than checking to see if the other departments are complying with standard procedures, the internal auditors present themselves in a non-threatening, helpful manner, asking: "What keeps you up at night—the problems you have trouble solving and where you need objective third-party advice?"

The maintenance department was a particular challenge. Wells says, "Typically the department is a different culture, a different environment. Clive had heard some rumblings that there were morale issues and that hits very high on our radar." Because maintenance crews work through the night at a hangar away from head office, they often feel isolated, both from their fellow employees and their families. "AIDS has been around for a zillion years: Aviation-Induced Divorce Syndrome," Tom Woods remarks wryly. Maintenance people say that increasing demands on them to fly to other bases doesn't help. "We don't want movie passes or parties as recognition," one says. "We'd appreciate the company offering to fly our families to us for a weekend when we're on the road for a couple of weeks."

After meeting with the new chief maintenance officer, Steve Ogle, the internal auditor spent three overnights in coveralls in the department and travelled to other bases to talk to mechanics. He learned that too much turnover in management had created communication problems. Ogle, who welcomed Wells's input, says: "There were a number of managers who didn't talk to each other and didn't work well as a team." For any five managers, "you had five airlines running, each of them doing their best, but without an overall vision of where to go and how to get there and the true cost of running a maintenance program." He had his four directors design an organizational chart based on the prospect of a hundred aircraft flying by 2010 and then went off with them for a three-day session to create an organizational chart that would define a department-wide vision and identify synergies among them. They now meet annually to review the vision and come up with a budget. Although Ogle says (humbly), "I'm not sure if I'm there yet," there have been great leaps in the reliability of airplanes and their operating performance—with financial results to match.

Throughout WestJet, the trend is to avoid getting mired in traditional ways of doing business, to be adaptable and open to streamlining operations. So Sandy Ruel, manager of specialty sales—which among many other things handles joint promotions with other travel-industry players—says, "Our rule is to keep it simple, not have mind-boggling paperwork." The deal with Signature Vacations, for example, is nailed down in a mere four pages covering such necessary details as flight delays and cancellations. "It's an agreement, not a contract. It's almost like a handshake." BeanLand's Derek Payne, the treasury director, had to view other airlines' agreements while consulting on a new stock-option plan with WestJet pilots. While an Air Canada agreement may be hundreds of pages, "our new five-year agreement is about five pages—simple and straightforward."

It was while meeting with the pilots that Payne theatrically defined the integral role productivity plays at the airline. As the pilots were insisting that productivity should be factored into their agreement, he piped up: "We have half the number of employees as Air Canada. We're all productive. The people in marketing and accounting are productive. I don't get paid extra because of my productivity." (He was obviously forgetting about the reservations agents' productivity-differential bonuses.) "I'm leaving until you quit talking about productivity," he said, walking out and slamming the door. He recalls that one pilot came to him afterward and said that his speech was "good ... it gave us a slap"—by being a positive reminder of the productive labour that most people at the airline try to perform routinely.

TO THE RESCUE

The fruits of planned productivity at WestJet surface most noticeably at times of stress, especially during what the industry calls irregular operations—IROPS. One of the more difficult IROPS happened in the summer of 2001 when the airline's dispatch and flight-planning computers shut down, essentially grounding its jets in Calgary and leaving the operations control centre (OCC) in the dark about incoming flights. "It was catastrophic—probably the worst day in WestJet's history," Mike Pritchard says, not counting the voluntary grounding of 1996. In March 1999, Michele Derry's department had smoothly taken over the dispatching of its own flights from Skyplan Services of Calgary, a flight-operations support provider (and has since grown from a staff of eight to about seventy-eight in early 2004).

During the Calgary IROP, Stan Davidson, a team leader of customer-service agents at the airport, was called back from interviewing CSA candidates at head office. He found nearly a thousand customers in lineups at the check-in counters. From noon till 2 the next morning, he was part of a hastily assembled super-team that tried to juggle the needs of stranded passengers. As he says, "There are only so many flights on Air Canada and so many hotels." For a party of five en route to a wedding on Vancouver Island, he secured seats on another airline and then booked them on a small carrier landing close to their destination. He managed to get another couple to a family funeral. Airports director Dale Tinevez and other directors and senior managers from headquarters soldiered on the firing line. "They were filling out vouchers, arranging for wheelchairs, taking people to hotels ... The culture states we work as a team, we're all equal, working in the trenches. It's one thing to say that's the vision, but it's another thing to actually see it working."

WestJet faced a similar test on April 4, 2003, when a late-winter blizzard was storming through Ontario and the entire lower floor of the airline's Calgary hangar lost power—disconnecting the suite of computer software that virtually runs its operations. This time, a team from the operations control centre and information-technology experts could take advantage of an upgraded computer system. As OCC manager Shane Harney reported in *Jet Lines*, the internal newsletter, they discovered that the hangar's second floor worked fine using the same network. The team moved four computers and monitors upstairs, where dispatchers could keep the airplanes flying seamlessly across Canada. Another team proceeded to a backup control centre, away from the hangar, that had never before been tested in a live situation. "OCC and IT were able to keep the operations moving for the entire day from our off-site location," Harney wrote. "Throughout the day, IT had sufficient time to troubleshoot the original problem and was able to restore the network ... Many of the pilots who had come into OCC, after flying most of the day, didn't even know there were problems until we filled them in on what had happened." He ended his article by thanking, by name, fifty-three WestJetters involved in the rescue operation.

The most significant recent IROP in aviation history was, of course, when transatlantic aircraft were diverted to Canada after terrorists attacked New York and Washington, DC, on September 11, 2001. Canadian airspace was soon shut down, throwing airports across the country into chaos. As Mike Pritchard says of his crew at the terminal, "This team was the shining star of this airport." Ethel Egglestone, who'd been with the airline from the beginning, was now training customer-service agents. WestJet's agents, along with their team leaders, stepped in to assist the American carriers at the airport that were being flooded with unexpected flights. "The CSAs weren't told to help the other airlines—they just did. And it just grew," she remembers. "Then everybody started helping. They helped re-accommodate guests, answer questions, and hand out meal vouchers, and then find hotel rooms. We were even behind the counters of the American airlines."

On a wall outside Mike Pritchard's office at the airport hangs a framed letter from James E. Goodwin, chair and CEO of United Airlines. He probably sent out other such letters after September 11, but this one is particularly cherished by the people of WestJet. Acknowledging the Canadians' efforts when a half-dozen of his airline's flights diverted to Calgary, he concludes: "WestJet employees provided tremendous support and assistance to our Calgary staff by meeting our flights, helping our customers and crews, and serving food and beverages. In so doing, they demonstrated the kinder, gentler side of mankind to individuals shaken by the day's events. Thank you for your kindness and compassion during a very difficult time for our company and our country."

And while WestJet's people were being nice to the Americans, their colleagues were positioning the airline to be the first in Canada to get up and fly with a full schedule in the traumatic week following 9/11.

10

KNOW WHEN TO SPREAD YOUR WINGS
Going public, expanding, competing

WestJet Pop Quiz: How many people named "Jennifer" does WestJet employ? How many positive e-mails did WestJet receive in 2002? How many guests did WestJet fly in 2002? How many resumés did WestJet receive in 2002? (*Answers:* 48; 7,000; 5.8 million; 49,000.)
  —Annual report, 2002

Was it *People* magazine or the annual report of a multi-million-dollar Canadian company? The cover had a newsstand-like bar code, cameo photos that included a young couple hugging, and punched-up sell lines: "Spotlight on the Stars of the Airline Industry ... The Unsung Heroes of WestJet." Inside were a letters column with customers gushing praise of the airline's employees; parodies of ads, including one showing Calgary comedian Jebb Fink, creator of TV's *An American in Canada*, with pretzels in his hair and ears ("On WestJet, even pretzels are fun"); and a fashion page spotlighting flight-crew uniforms, which quoted purchasing agent Arlana Zeyha in mock-designer prose ("Blue really took off last year. It's the focus of our whole line, allowing for a few symbolic white and teal accents").

WestJet's founders appeared on the board of directors' page at the back in unstatesmanlike poses: Clive Beddoe unloading a dishwasher, Don Bell with a giant fish biting his ear, Mark Hill shouldering a driftwood log—"taking stock of materials before sadly realizing the Spruce Goose II would never be economical on short-haul flights."

Oh, there were solid accounts of the airline's results in 2002—disguised as magazine articles—and the requisite management and auditor reports to shareholders. But in its snazzy, colourful design and satirical tone, the *WestJet Annual Report/The Magazine* was as far from a traditional financial-reporting document as the airline itself is from a conventional carrier like Air Canada. Hell, you wanted to read the damn thing.

Well before WestJet offered itself to the wider world of investors and became a public company in 1999, it was making itself a transparent corporation. The year before, it distributed an annual report that spelled out all the key fiscal details of the privately held enterprise, including a 63-per-cent leap in revenues to $125.9 million. The '99 report, the first after its initial public offering (IPO), presented the playfulness of the company to bemused shareholders with a paper airplane on punch-out cardboard—one of the new-generation 737-700 series of jets WestJet would be buying from Boeing in 2000 (at $35 million US apiece). The next two annual reports were much more conservative. Perhaps that's because everyone was too busy sending the airline into the financial stratosphere while moving into expanded new digs and, after long deliberation—not to mention a few years' public denial—heading east.

During 2000, WestJet consolidated four cramped offices into a new four-storey building of 67,000 square feet near Deerfoot Trail and McKnight Boulevard, a brief drive from the Calgary airport. The old headquarters had become so congested that the Big Shots had to move out to a satellite office. The company leased the larger space for ten years, expecting to move again into a head office it plans to build next door to its own $20-million hangar near the north end of the airport. Opened in 2001, the sprawling hangar of 220,000 square feet can hold six 737s at a time for maintenance. It houses sheet-metal and engine shops, an aircraft-cabin trainer, two flight simulators, the operations control centre, parts stores, and a good chunk of the staff that soon began overflowing the new office space.

All this expansion was being fuelled by holders of the newly public shares. WestJet had always intended to go public. "We knew we needed to access the public markets to give meaning to the stock-purchase plans for the employees and to have access to the capital," says director Ron Greene. In 1997, when Clive Beddoe had asked Sandy Campbell to become chief financial officer, replacing Don Clark, Campbell said, "I suspect you'll want somebody with more experience; I've never taken a company public before."

"Neither have I," Beddoe replied.

And at first nobody on staff at WestJet knew much about doing an IPO. Fortunately, some of the corporate directors and lawyer Daryl Fridhandler did. In 1998, the airline became a reporting issuer—a corporation that issues or has outstanding securities held by the public—and was automatically subject to the disclosure requirements of securities watchdogs. Taking this step a year in advance of the IPO was Fridhandler's idea. He knew that Clive Beddoe in particular hoped previous investors in the private company—including the employees—could trade shares immediately after WestJet went on the stock market. But in Canada, shares issued before an initial public offering are under an embargo that prevents them from being traded for a full year after the IPO. Unless, that is, the company becomes a reporting issuer earlier, as Fridhandler had the airline do by completing a prospectus with five provincial securities commissions in the west and Ontario. By the time WestJet shares began trading, all the loyal investors in the private company could legally buy and sell the stock.

Mark Hill, who by now knew the aviation industry so well, was liaising with the lawyer on the prospectus. At the time, Steve Smith was WestJet's new president and, along with Clive Beddoe, CFO Sandy Campbell, and Hill, he was part of the team that took the IPO on the road to sell to big investors. "He had just started in earnest in March and he's talking about WestJet as if it was his," Hill recollects a little irately more than four years later. "In Boston and New York, he told me, 'We really don't need you.' And I thought, *Screw you, I've worked too hard and too long to be pushed aside.*"

Liaising with Hill was a suddenly empowered Derek Payne, who'd joined WestJet in his late twenties with a bachelor of commerce from the University of Saskatchewan and six years' articling with KPMG in Calgary. Although his role was supposed to be cash management, his first job (a month after starting) was coordinating the IPO within the company as the "project worrier—making sure the information was accurate, educating employees on their shares and stock options." He says WestJet did an initial public offering in its own distinctive way: "In most companies, the rich get richer [in an IPO]. A certain number of shares are set aside for the company and 99 per cent of the time, they go to the president and the rest of the executive." His task was to ask employees if they were interested in buying shares of the public company. There was such response that each WestJetter, no matter the rank, got only 190 shares worth

$1,900. "Clive Beddoe got 190 shares; Lou Oneski, our accounts-payable clerk, got 190 shares—which sends a tremendous message to the pilots and mechanics, labour groups who traditionally look skeptically at management."

PR and communications director Siobhan Vinish found the process of going public a fascinating exercise in contrasting cultures between a Calgary-based company and the eastern banking establishment that she and Beddoe met with in Toronto: "The bank had such a hard time letting us be who we were. I remember us being in this great big boardroom and they were trying to talk to us about the presentation Clive is giving. They were trying to convince us to be more structured and more bank-like. That is not who we are and I remember thinking, *You guys are trying to put us into this box, because that's what all your other people do.* We *are* that 'aw shucks, down-home' kind of company. We have a good product, so we've got a great story to tell about who we are and the values by which we live. Yes, of course, you have to back that up with good financial results and a strong management team ... And I think Toronto—downtown Toronto—had a challenge with that."

When the airline went on the Toronto Stock Exchange in mid-July, 1999, shares began trading at $10 and rose about $3 within the first couple of weeks. (Derek Payne bought himself a camcorder with the profit on his shares.) The stock has since split twice: the first three-for-two split was in May 2000 and the second two years later. Recently, it was trading around $30. Obviously, WestJet's style hasn't bothered most investors, including the Ontario Teachers' Pension Fund, which in 2003 agreed to put in another $100 million, and Fidelity Management, the American mutual-fund giant, which became the company's second-largest shareholder.

As Sandy Campbell says, the IPO wasn't much of a culture shock for the company because WestJet had prepared itself "by thinking well down the road and having the vision of what you're going to be when you grow up. All the way along, because Ontario Teachers' have been on the board, we looked and felt and acted like a public company." Some of his people, however, faced what accounting director Janice Paget calls "a huge change in the way we did the compliance and disclosure side of quarter-ends and year-ends."

Jeremy Forrest became the company's investor-communications specialist, spending 70 per cent of his time writing press releases and financial reports and answering investors' questions ("Why has the price gone down? I don't want any corporate BS; I want the truth"). With an English degree and a minor in religious studies from the University of Calgary, he had to get up to speed on finance. "Sandy would sit in my cubicle and go through things. I felt completely comfortable saying I didn't understand. He'd draw a chart and make analogies ... He *lives* empowerment." Forrest wrote much of the magazine-like 2002 annual report with communications coordinator Sarah Deveau. Then he and another twenty-six-year-old, accountant Shirley Saputra, went over the financials and signed off on them the day before the report went to press—"We didn't have to call Siobhan or Sandy." But cooler heads had won out in editing the text, rejecting a satire of a Volkswagen ad and a parodic cover title reading "People" instead of "annual report." Still, enough of the humour survived to present WestJet as a folksy, accessible company to its shareholders. No wonder its investor-relations program—which blends BeanLand's financial acumen and the marketing department's PR skills—won an award in 2003 from the British-based *IR Magazine* as the Canadian company having the best communications with the retail market.

GO EAST, YOUNG AIRLINE

Timing is everything in takeoff. You have to be going fast enough, create enough forward thrust to overcome the drag, to become airborne. With aircraft, taking off safely is a matter of aerodynamics and experience. With airlines, it's more mathematics and instinct. For WestJet, 1999 was the year the numbers came together and it felt right, seemed to make sense, to leave the friendly skies of western Canada and fly into the east. "The impending bankruptcy of Canadian Airlines was the most significant trigger for us to go east," Clive Beddoe says. When his company announced its plan that December, the share price hit a record $18.65.

Expansion was a group decision that drew on the abilities of the four founders, from the executive skills of Clive Beddoe to the intelligence-gathering strategies of Mark Hill. That, and the collective wisdom of the corporate directors (who now included Donald A. MacDonald, president of Sanjel Corporation, the largest privately owned Canadian oilfield-services company, and Larry Pollock, president and CEO of Canadian Western Bank and Canadian Western Trust). And especially the market analysis and schedule planning of Bill Lamberton and pivotal people such as Brenda Trockstad, director of revenue and schedule—"the core operation of the marketing product"—and revenue manager Ben Druce, who helps determine the various fares for each destination (the closer to time of departure, the higher the fare).

As they had in selecting Abbotsford, British Columbia, the company's planners looked for an airport with just the right balance. It had to be close enough to a major population base, yet just off-centre enough to be much less crowded with rival aircraft—to facilitate fast turnarounds—and cheaper in landing fees and terminal expenses to maintain the all-important competitive cost advantage. An obvious choice was Hamilton International Airport, less than an hour's drive from Toronto with its jammed and expensive Pearson International Airport. Hamilton, where fees and charges were about 40 per cent of Toronto's, was ideally placed to draw on the enormous population pool of the Golden Horseshoe encompassing the Ontario capital and the Niagara Peninsula.

All this was obvious to WestJet perhaps, but the year before it began using the airport, a mere 22,000 passengers had passed through the terminal. The year after, there were ten times as many. Unlike Abbotsford, which remains a spoke to Calgary's hub, the Steel City's airfield became the principal eastern base. There were some cultural challenges to face: incorporating the French language into WestJet's operations and exporting the corporate values to staff in distant cities. The airline launched modestly with two daily flights to Thunder Bay, Ontario, and Winnipeg (its regular one-way fare to Winnipeg was $125, about a sixth of Air Canada's). By year's end, it was flying to Moncton and Ottawa. In 2001, adding service to Sudbury and Sault Ste. Marie, the airline put half a million passengers through Hamilton and a million by the end of 2003—when it threw a party in the form of a western country fair and barn dance to announce its new maintenance hangar there. By then, the airport had become the ninth-largest in Canada and some observers say the fastest-growing in North America—until early 2004, when WestJet finally bowed to Toronto's larger population and decided to move 60 per cent of its flights from Hamilton to Pearson.

An airline intent on expanding has to analyze external circumstances just as a pilot does in deciding when to take off (how strong the winds are, how heavy the snow).

WestJet had researched what the business climate was like, what the competition was doing at a time—the cusp of the twenty-first century—when a lot was happening in the Canadian industry.

The airline had announced its eastern expansion in December 1999, the same month thirty-nine-year-old Robert Milton took over Air Canada's controls from Lamar Durrett. He had to conduct a very public political as well as legal fight to prevent a hostile takeover by investor Gerry Schwartz, the CEO of Onex Corporation, who wanted to merge the airline with Canadian Airlines. Milton won, but then merged the two carriers anyway, infuriating federal politicians and consumers with the monopolistic move. The acquisition signalled Canadian's end and strengthened Air Canada's resolve to create its own discount airline in direct competition with WestJet. It was obviously time for the westerners to wing eastward. Initially, Air Canada had intended to locate its first low-fare subsidiary in Hamilton, but in 2001 its Tango fleet began operating out of Toronto to certain high-traffic Canadian and US destinations.

A year later, Steve Smith became the founding president of Air Canada's Zip, based in Calgary. Taking dead aim at WestJet's western operation, the no-frills Zip took over its parent's mainline, full-service business in the west. While Beddoe and his colleagues had obvious concerns about these forays into their airspace, they also knew that the unionized Red Baron's cost structures were prohibitively high. Even with agreed-upon pay cuts, its pilots flying for the discount carriers would be earning a base salary twice WestJet's. "You don't lower the cost of your operation by re-painting airplanes," Beddoe says today. "You can put seats out there for $40 between Calgary and Abbotsford, but they're losing a fortune doing it." (Later, Air Canada set up

Jazz, an eastern regional subsidiary flying turboprop aircraft into small communities, which used pilots from another union who were paid less.) WestJetters were also heartened by the reminder, in a flattering *Canadian Business* article by industry observer Peter Verburg, that "every attempt by a major scheduled airline in the US to launch a no-frills, low-fare alternative has been a disaster."

More ominous at first was the profusion of other airlines crowding the marketplace. With Canadian's demise, Canada 3000, which had begun as a charter business in 1998, was now intending to become the number-two national carrier with scheduled domestic flights. Meanwhile, Royal Airlines, a spinoff of another charter operation, had begun low-fare scheduled service to Winnipeg and four eastern cities, offering frills such as meals; in 2001, it would double its flights to include Vancouver, Calgary, and Edmonton. And a new discount company called CanJet Airlines had threatened to base itself in Hamilton, but just before WestJet's launch there, owner Ken Rowe decided to fly out of Halifax to serve Central and Atlantic Canada. As it soon turned out, Canada 3000 bought both CanJet and Royal and then went bankrupt in late 2001. Among many other things, it was a victim of the travel drought following the terrorist attacks on the United States that September and of strong competition from Tango. (To underline just how complex the Canadian industry is, CanJet has since resurfaced in Halifax and the former head of Royal Airlines has started the low-fare Jetsgo based in Quebec—which offers such promotions as Loonie Sundays when hundreds of seats go for a dollar).

To counter the competition and promote WestJet's new base as a gateway in early 2000, Bill Lamberton, Siobhan Vinish, and their marketing people had begun a print ad campaign designed by Calgary's Creative Intelligence

Agency. It was aimed at the Golden Horseshoe cities of St. Catharines, Guelph, and Waterloo, as well as Hamilton, trying to convince travellers of the underused airport's advantages, such as proximity and low-cost parking. A key public-relations element was personal contact with travel agents—taking them on familiarization trips to show how convenient the Hamilton field was while entertaining them with a night on the town and a drive to Niagara Falls. The marketers also worked closely with the airport authority and business newspapers in the Niagara region to publicize the hometown airport—"fly locally." Later, the airline ran a series of newspaper banner ads in Hamilton and key western cities to heighten awareness of its schedule and frequency. Some advertised specific flights, such as the "Zzzave while you sleep" promotion of WestJet's first night-flight from Calgary to Hamilton. Occasionally the price-point print campaign was bolstered with billboards and radio spots—as opposed to expensive TV commercials. As its competitors kept running into turbulence, WestJet got enough free publicity on television as the western hero in the Canadian airline industry.

The airline's geographic roots seemed to help, not hinder its reputation. Although the company had formally reserved the name EastJet, it never seriously considered changing its moniker to accommodate the folks from the right-hand side of the map. "I don't think we've ever had a problem flying in Eastern Canada," says co-founder Tim Morgan, sitting in his office in Calgary. "If you go way down east to the Rock and stand in St. John's and look in this direction, which way are you looking? *West*. There you go, right?"

Not all the easteners were crazy about The Little Airline that Did. Air Canada was determined to beat the upstart at its own game. At the time WestJet had decided to fly from

Hamilton to Moncton in 2000, with a one-way fare of $129, it thought the senior carrier was reducing service in the Maritimes. "At that point," says Mark Hill, "they were rationalizing their schedule and we could see that they would cut back flights by 10 to 15 per cent, which would make our numbers there even better. We knew they'd *match* us, but they undercut and flew three 737s instead of F-28s and increased their capacity by about 50 per cent." Air Canada reduced its fares to and from Moncton by 75 per cent. Robert Milton said during a parliamentary hearing, "WestJet is now going to have to compete, but we will not behave in a predatory fashion." WestJet, claiming the prices were predatory, estimated that on the Moncton-Toronto route, its competitor was losing at least seven cents per available seat mile. Obviously Goliath wanted to force the little Davids out of the Maritimes, where it had enjoyed a near-monopoly for so long. *We gotcha!* Hill thought. "This was so over-the-top, it was a no-brainer. We'd been talking to the Competition Bureau for years and had a relationship with them."

Both WestJet and CanJet (before it was bought by Canada 3000) complained to Industry Canada's Competition Bureau. Its commissioner took the case to the bureau's Competition Tribunal, alleging anti-competitive conduct by a dominant firm to substantially lessen competition. Hill oversaw WestJet's project to intervene as Alicia Quesnel and Dan McDonald, two lawyers from Burnet, Duckworth & Palmer, worked with data being coordinated at the airline by a new kid on the block.

Scott Butler, with a commerce degree from the University of Victoria, came to work at WestJet in late 1999 as an internal auditor—for which he had no experience— and was soon interested in applying for a position that had been posted for a corporate planner. "The problem was," he says, "it asked for industry experience and an MBA—and I wasn't even close." He questioned Sandy Campbell why the qualifications were so lofty. "Don't let that bother you," the CFO said, enigmatically. What he didn't say was that Steve Smith had made those demands. And a little while later, on the day after the president was let go, Campbell asked Butler if he was still interested in the job. The twenty-five-year-old, six-foot-five string bean, who looked like a college basketball player, wound up labouring full-time on the Air Canada competition file. As he interprets the issue, "a dominant airline cannot operate at fares below the avoidable costs of providing the service." The avoidable costs of a flight are those, such as staff wages or jet fuel, that Air Canada would not have incurred if there was no flight.

Butler became a corporate presence in Ottawa, calling on the expertise of transportation consultant David Gillen of Waterloo's Wilfrid Laurier University, along with the legal team. Air Canada's legal consultant was Lawson Hunter, the former head of the Competition Bureau. Three years would pass between the time WestJet filed its original complaint and the tribunal rendered the first phase of its judgment. While saying that a second phase would be held to determine whether Air Canada had abused its dominant position, the tribunal did conclude that the airline had cut fares below its costs on two routes in the Maritimes, increasing capacity while not covering avoidable costs.

"It was an exceptional win," Scott Butler says. "What was shocking about it was that on every single point where the tribunal had to make a decision, they sided with the commissioner and WestJet. They never sided with Air Canada. We couldn't have expected better results. One little tidbit that emerged was that the Air Canada avoidable costs per available-seat-mile were 23.1 cents from Toronto to

Moncton. For us to fly Hamilton to Moncton, it costs about 9 cents per seat mile."

During 2003, WestJet began flying to Halifax and to St. John's and Gander, Newfoundland.

THE LOAN ARRANGERS

Confident because of the accelerating growth across the country, the company was encouraged to begin replacing its maturing fleet. The average age of its second-hand Boeing 737s was twenty years. By the end of 2000, WestJet had twenty-two of the 200 series, which it intended to retire within eight years. Now it would order its first-ever new aircraft and they would be the sexy next-generation 737-700s. Flying faster, farther, and more quietly than their predecessors, the 136-seat jet airliners would bring better fuel efficiencies and lower maintenance costs. But they'd come at a steep price: buying twenty-six of them from Boeing (with an option to acquire up to forty more) and leasing another ten through GE Capital Aviation Services would require $1.2-billion ($744 million US) in loan guarantees from the US government's Export-Import Bank. Ex-Im, for short, is the official credit agency supporting American companies with export financing.

In WestJet's headquarters, the deal became known as the Big Honkin' Loan Application. Treasury director Derek Payne, who'd begun coordinating the IPOs with only thirty days' experience at the company, was now handling the loan negotiation with no background in debt financing. "We had a beauty contest," he says. "Twelve banks came in and presented. Ultimately we selected ING. They were North American and European lenders. There were no Canadian institutions involved in the process because of their lack of experience in export finance with Ex-Im and typically they

216

do not like airline risk. ING won primarily due to the simplicity of its deal, and the fact that they provided the best solution to meet our objective of providing Canadian fixed-rate funding so that we could eliminate foreign-exchange and interest-rate risk."

A major problem loomed, however. "Because the US government is guaranteeing the loans, they have a structure to protect themselves, but it didn't work with our tax laws. The primary issue was that under a finance lease—Ex-Im's preferred structure—WestJet would not be able to claim tax depreciation on the assets. As WestJet is profitable and therefore cash-taxable, the writeoffs in the initial years are valuable to the company." The airline's negotiators wasted two and a half months trying to squeeze Ex-Im's square peg into the round hole of Canadian taxation rules. "The pressure's immense to make it work from our tax perspective," recounts Dino De Luca of Burnet, Duckworth & Palmer, "but we find we're not getting their attention on matters important to WestJet." A little desperately, with only eight days before the first of the new aircraft were to arrive, the Canadians decided to send a team of six to the Export-Import Bank's headquarters in Washington, DC.

It was to be the first time the two sides would meet face to face. With his fresh, unlined face, the twenty-eight-year-old Payne looked almost a decade younger. WestJet's in-house legal counsel, Shawn Christiansen, was a mere two years older than him. Of the external lawyers, the lean, youthful-looking DeLuca was the oldest at forty-one. His crew included a thirty-five-year-old banking lawyer, a tax lawyer a year younger, and an articling student in his twenties.

When they arrived at Holland & Knight LLP—the bank's legal counsel located three blocks from the White House—the WestJet team realized for the first time just how blue-chip the firm was. "Their premises were gorgeous, including the big-screen TV with CNN in the lobby, two receptionists meeting you, security everywhere, two computers in a big boardroom, and each of their lawyers had a laptop," DeLuca says. "We meet Robert Wray, a sixty-five-year-old senior lawyer affectionately known to some of us as Grandpa afterwards." In terms of youthfulness, Payne says, "we are not who the Ex-Im bank traditionally deals with. When they came in, you could see they were surprised."

The bank's legal team insisted again that the concerns WestJet's people had raised—about melding Canada's tax laws with Ex-Im's financing structure—were not of serious consequence, and suggested they go back and talk to their Canadian tax consultants again. But the two parties kept discussing the taxation issue until 2 or 3 the next morning, with the Americans caucusing in an adjacent boardroom. "We didn't gain their respect, but we did get their attention," De Luca says. "Then they did a 180-degree switch and said, 'We've got some serious issues and we'll have to deal with this. Robert Wray of Ex-Im is going to be calling Sandy [Campbell] and discussing this with him.' This is a serious blow to our team. Sandy basically has said whatever we decide, he supports. We're more like business partners and get treated like a full member of the WestJet team."

During one of the time-outs, DeLuca and an exasperated Payne discussed strategy and then confronted the bank's lawyers. "First of all," Payne told them, "we're offended. I apologize if we're not what you were expecting. You guys have an issue that you're not willing to share with us. We're sitting here fifteen feet away from you—waiting."

DeLuca pitched in: "When we walked in these doors today, we talked about respect and working to get a deal done. At the end of the day, you're saying you decided to call Sandy Campbell. We are the people who will negotiate and we will decide if and when Sandy is called. This isn't the B Team and if we're not successful, they'll then send in the A Team. This *is* the A Team."

It was a defining moment, he remembers. "They weren't taking us lightly any more. They left the boardroom and caucused together and came back to explain all their issues." The bank was worried about accessing security, how it would get the aircraft back in a timely manner if WestJet defaulted on the loan. "The solution was to treat the lender ING and Ex-Im Bank differently, splitting fine hairs to fit within the tax rules and give them the security they wanted. You could see the eyes of Louis Emery, the in-house lawyer for Ex-Im, light up, thinking these guys have got a good solution." Emery confessed later that the first time the WestJet team left the boardroom, he turned to his colleagues and asked, "Where are their dads?"

For the next week, the Canadians closeted themselves in Washington, working on the details of the deal and having only one dinner and a midnight pizza outside the building. "The day before closing," DeLuca says, "Robert Wray heard Sandy and Clive were flying in and said, 'Hey, I guess the B Team's coming in.'"

Flying home to Calgary, a delighted Campbell and Beddoe insisted that DeLuca and his articling student use the WestJet executives' business-class seats and that they'd sit in economy.

TAKING TORONTO

WestJet has a hard-headed attitude to the routes it chooses to fly. This is summed up in a common saying the marketing and sales people employ with the smaller communities: "Use us or lose us." If a new destination fails to support a route with traffic after a reasonable test period, the airline will simply drop it from the schedule. In 2003, for example, WestJet ended service to Sudbury and Sault Ste. Marie, Ontario, after abandoning Thompson, Manitoba, the previous autumn—all because of falling demand. That's a luxury Air Canada, as the mainline national carrier, hasn't always allowed itself—although it will charge steep fares on non-competitive routes. (WestJet also occasionally offers a service called Limited Addition, which uses aircraft that have arrived earlier than expected and can't fit into the core schedule. In that case, it has provided regular flights to smaller centres such as Brandon, Manitoba, for a publicly defined time period.)

Pick and choose as it does, WestJet could not avoid one destination: the GTA, the Greater Toronto Area—YYZ, in airport code. Despite its pioneering of Hamilton's low-cost airport, the company couldn't really put off flying into Toronto's pricey Pearson International forever. Yes, it was the nation's busiest, the heart of Air Canada's operation, but many discount passengers wanted the convenience of landing within the largest Canadian centre, the financial capital.

With service scheduled to start in May 2002, how would the airline pitch itself to the city's media-savvy inhabitants? Clive Beddoe believes in word of mouth, "which is ten times more powerful than any advertising campaign." But what about an aggressive campaign that can provoke such talk on the street? In an interview with *Marketing*, Bill Lamberton had said WestJet would continue its traditional, low-key advertising and PR strategy in Toronto, one that emphasizes price and customer service. But he didn't tip his hand that the game plan would also include an eight-week "station-domination" campaign that autumn—underlining the airline's distinctive brand—at Toronto's downtown transportation hub, Union Station.

Handling more people a year than all three terminals at Pearson, the station sees 160,000-plus passengers each business day. GO commuter trains, the GO bus service, the subway, and VIA Rail, Amtrak, and Ontario Northland Railways all use it. CIA's (Creative Intelligence Agency) idea (which had been pioneered by other advertisers) was to blanket every available surface at the station—conventional poster frames and non-traditional media space such as columns, ceilings, stair risers, wallscapes, soffits, and floors. The airline was the only brand name in the public areas of the building. Over two days, about twenty off-duty WestJet staff handed out bags of airline snacks and 10,000 free copies of the tabloid *Toronto Sun* with a wraparound cover about WestJet and a teaser about an upcoming airline contest in the newspaper (grand prize: trips for four to three cities to either follow the Maple Leafs or indulge in spas).

Dean McKenzie and Barry Anderson of Creative Intelligence Agency designed a cheeky campaign that had the westerners gently poking fun at Toronto: "We're new in town—don't freak out if we smile at you"; "Finally something affordable in Toronto"; and ("dealing with the whole Toronto angst thing," McKenzie says) "It's okay to bring baggage into this relationship." One tag line took on the competition: "Returning your opinion of airlines to the upright position." Another pointed out the effect of low-fare travel: "We apologize in advance for the extra in-law visits." And a poster of a jet forming a grin from its contrails read: "Smile. You'll be on WestJet soon."

And they were. "Our Toronto services were launched with great success from the point of sales, right from the get-go," Lamberton said. WestJet was soon offering seven flights a day with non-stop service from Pearson to Calgary, Edmonton, and Vancouver. Business travellers—who often book higher-priced last-minute fares—were attracted by the convenience of flying in and out of Toronto.

DEFINING THE DIFFERENCES

Toronto was now tied into the WestJet network; Montreal followed in April 2003. The next challenge was to grow from the market-stimulation model—introducing new customers, the infrequent travellers, to the airline—and try to attract those regular flyers who were still faithful to Air Canada. "For seven years we've been market stimulators ... by offering lower fares," Lamberton was saying in mid-2003. "We got people out of their cars and trucks on the short haul. Then we grew into transcontinental market stimulation and that's working pretty well. But certainly in Western Canada, there's not much more that we can stimulate; they're getting used to low fares. We're mass marketers, but now we have to go and get people to switch ... So our challenge now is more ridership. We're going to have to ask people that have been loyal to another airline: 'Have you looked at WestJet lately? We've got more schedules now, we fly fourteen times a day between Calgary and Vancouver.' We haven't targeted anybody, but we might want to make the business traveller aware of that.

"We've been in the air with the Boeing 737-700s for two years. And we're now adding product differentiation ... We've never had a loyalty program. But now you can get Air Miles, which is in 65 per cent of households in Canada. So we'll use that as an added value to our product. And by gosh, you know what? Satellite TV is coming on our airplanes ... We've studied the heck out of it, and we're seeing that it's going to increase ridership and, by golly, it's not going to increase our costs very much."

Satellite television at every seat is the first real on-board frill the airline has ever offered. In a deal with LiveTV, a subsidiary of David Neeleman's innovative JetBlue in New York, it planned to install personal video screens on forty of its new aircraft during 2004—at a cost of half a million dollars apiece. At first the service will be free, but WestJet may later charge a nominal fee. Bell ExpressVu will offer a choice of twenty-four live satellite channels, including the major North American networks and some specialty offerings such as business and cartoon channels.

The next-generation 737-700s have cabins that are slightly more luxurious and roomier than the 200s. While Boeing built the guts of the airplanes in Renton, Washington, and Wichita, Kansas, other suppliers furnished components such as seats, galleys, and even engines. A WestJet team headed by chief maintenance officer Steve Ogle decided on the various options and their vendors. They chose leather seat covers, which are initially expensive but longer-wearing and far less costly to keep up. The airline was the first to install stowage compartments called Big Bins in 737 cabins. They're about 60-per-cent larger than conventional luggage bins, holding more than thirty extra carry-on bags. And WestJet is removing four rows of seats and unused lockers and bulkheads in all new aircraft to provide more legroom on the longer flights it's making. The new seat pitch (measured from the back of a seat to the back of the one behind it) is 81.28 centimetres instead of their usual 78.74. Although it doesn't sound like much, every passenger knows that in the confines of an airline cabin, every centimetre counts.

The 700s can reach an altitude of 41,000 feet, higher than any other 737—well above bad weather and the crowded skies over densely populated areas. They fly at speeds up to 860 kilometres an hour, cruising at 770 kmh compared with the 690 of older 737s. They can stay aloft for 6,038 kilometres,

1,600 more than their predecessors—increasing the number of city pairs the airline can link. Not only do they have lower maintenance costs, but they also offer a 30-per-cent improvement in fuel economy. Stu McLean led the company into becoming North America's first airline to install a further economy measure on the aircraft: blended winglet technology. The winglets are sleek-looking wing-tip extensions that curve upward nearly 2.4 metres to reduce drag—which increases flight range, extends engine life, and saves about 4 per cent more in fuel. That was one percentage point less than WestJet had expected, but any little edge in lowered costs was welcome.

But the most important refinements of the 700-series jets are the safety enhancements they incorporate. A Heads-Up guidance system will let pilots concentrate on a single critical area for visual and flight data information during low-visibility and night-time approaches. And the new aircraft are fitted out for a high-tech, $2-million-plus option that WestJet plans to have operating during 2004: required navigation performance (RNP). Again, it will be the first Canadian carrier to use the RNP instrument-approach computerized system, which essentially gives an aircraft the ability to know its own position exactly, independent of air-traffic control. Sometimes called "pinpoint navigation," RNP uses a combination of global-positioning and inertial-navigation systems to create a line for the airplane to follow on a precise three-dimensional course. "It's a revolutionary way of getting an airplane into a difficult airport at night in a snowstorm," Stu McLean says. Officials from Transport Canada and Nav Canada, the private company operating the national civil air-navigation service, came from Ottawa in autumn 2003 and were impressed after flying aboard one of WestJet's RNP-equipped 737s into the tricky Kelowna airport.

Tim Morgan says flight-operations director Bruce Flodstedt is "the guy that put all the equipment in the airplane. He has got another smart guy working for him, Dave Hathaway. He's telling Boeing about their own products. And he's the RNP wizard. We can physically fly that airplane to any airport—anywhere in the world, for that matter—down to 250 feet without any [landing-system] equipment on the ground. So that gives us one tremendous operating advantage in any weather." Another safety device being considered is the use of satellite communications for instant phone calls to crews on long-haul flights, rather than depending on VHF radio and faxes.

Bruce Flodstedt says, "Buying those new airplanes was the biggest challenge in our eight years. None of us had any experience with the new generation of aircraft with their 'glass cockpit'—screens and computers rather than the old dials and gauges." After a WestJet team of pilots trained at Continental Air Lines in Houston ("For these poor buggers, it was hell at 2 in the morning, going through the crash course"), they designed a training regimen with the help of a retired Continental pilot. "I believe we have one of the best training programs in the industry," Flodstedt boasts.

WestJet's new penchant for product differentiation (which should continue despite Lamberton's parting from the company) starts at the check-in counters at the airport. In December 2002, it moved into the Calgary airport's spiffy new D wing, which offered the airline about 60-per-cent more room. As a latecomer to the industry, says Dale Mitchell, WestJet's terminal support manager, "We worked in every piece of this airport and kept getting what was leftover space—and then finally moved into something brand-new and shining with all these bells and whistles. It makes Clive cringe when he comes in here because he feels it could be spent better building a Quonset hut and charging our guests less."

One of the bells and whistles WestJet introduced was Canada's first dual over-the-wing bridge, at the airport's Gate D40, which allows simultaneous front and rear boarding and deplaning of an airplane. A new kind of bridge, designed with advanced sensors by DEW Engineering and Development of Ottawa, extends over the wing to the rear door. "We use Gate 40 if an aircraft's coming in late because we know we can make up time," says Mitchell. "If the airplane's not in the air, it's not making money."

Mitchell, who was the airline's operations team leader for the airport extension, embraces the new with the exhilaration of a kid on his first plane trip. "We still have not lost that thirst for innovation. We have not become complacent." But as an industry veteran of thirty-six years, twenty-seven of them with Air Canada, he knows that WestJet's people-first culture is far more important than any fancy technology.

11

HAVE A FUN FLIGHT
Pleasing customers—and employees

WestJet ... the official airline of Fun.
  —Slogan on a WestJet call-waiting message

If you're not having fun, you're fired.
  —Clive Beddoe, only half-kidding

For flight attendant Trevor Arnold, it was shaping up to be the flight to hell. A ramp agent was doing a pushback—propelling a jet backwards with a push-tug vehicle—near the gate at the airport in Calgary one day in 2002. WestJet had leased the 737-800 briefly from Delta Air Lines. As the aircraft was backed up, its rear wing clipped a commissary truck parked behind it. The truck tipped over and the driver dived into the airplane to save himself. The damage was an estimated $400,000. And a planeload of passengers were badly delayed on their flight to Toronto. Arnold, a freckle-faced redhead from Lethbridge, Alberta, was off duty, visiting his girlfriend, a reservations agent working in the head office near the airport. It was just after 6 p.m. Wrong place, wrong time. The accident had left the airline's flight crew shorthanded; could he be ready to fly as lead attendant by 8?

Tre Arnold has a big laugh and an easy demeanour, but he was startled by the request. He was in street clothes, and there was no time to change at home; what the heck would he wear on board? The crew room had a stock of used uniforms, where the best-fitting shirt didn't smell so good. The largest pants had a 32 waist and he was a 36. He settled for a pair that had the right leg length and, to hide the 4-inch gap around his middle, donned a tall man's WestJet vest he'd burgled from a team leader's office. Because the airline's standard was black shiny shoes, not hiking boots, he located a pair of light-grey canvas shoes and coloured them with a black felt marker as the crew bus took him to the airplane.

Introducing himself on the microphone, he tried to distract the obviously annoyed passengers by telling a concocted story in a Gaelic accent about his Irish father and Scottish mother driving home from a pub. Dad is weaving and crashing the car until a constable stops him to ask if he's aware that his wife fell out three intersections back. "My father breathed a sigh of relief and said, 'Thank you, officer—for the last ten minutes, I could have sworn I'd gone deaf.'" For the safety demonstration, Arnold switched to a Scottish burr to talk about "this great big haggis in the sky" and then pointed out a flight attendant in the cabin who he claimed had ditched him. "I hope this will rekindle the flame," he said and sang in a nice tenor, "You've got that lovin' feeling."

By that time, the ill-dressed, joke-cracking, ballad-warbling attendant had broken down the passengers' defences. As they were landing in Toronto, he was on the mike again, asking if they wanted any assistance—"or just want a hug from a good-looking redhead." Most of them were smiling as they deplaned, thanking him despite the delayed flight—"and I even got hugs from chicks."

Low fares are one thing, but there, in the sure-can-do philosophy and relaxed good humour of Tre Arnold, is the publicly perceived essence of WestJet. Since 1996—in vivid contrast to the too-often grudging and even cranky service of Air Canada—it has wafted through the Canadian airline industry like a sweet, refreshing spring breeze across the prairie. The jokes and games, no matter how corny or relentlessly downhome they are, seem a welcome improvement over the solemnity of the usual flying experience. The smiles and personal attention, a little surprising and perhaps intrusive at times, are so much more comforting than a cattle-car-in-the-air mentality that fails to distinguish between individual flyers.

David Granirer is a Vancouver psychotherapist and a lecturer on humour in the workplace who also performs as a stand-up comic. Flying on WestJet for the first time, he heard the flight crew turn the safety announcement into a comedy routine. "Upon inquiring, I learned that the CEO of WestJet believes that work should be fun ... And the payoff? All WestJet employees I spoke to said how much they loved working for the company because it was so much fun. And their enjoyment was reflected in great service to the passengers. They expressed their appreciation of the humour and their intention to continue flying WestJet."

The feeling of fun, the apparent concern, the obvious empowerment—all spring from employees who carry an authentic sense of ownership in their company. They realize that their positive image and helpful practices will reflect directly in their profit-sharing allotments and share-purchase gains. Since the airline's start-up, when WestJetters knew their regular incomes would never match their major rivals', they've received $56 million from the profits of an enterprise that has not yet had an unprofitable quarter. Their shares, traded at $10 in the launch of the public company, have split twice and hit a price six times the opening figure—or more than $60 when adjusted for the splits (with another split scheduled in 2004).

No wonder they seem to be having so much fun. Who hasn't flown WestJet and heard gags like: "Someone phoned our line the other day and asked how long it took to fly from Vancouver to Calgary. When the customer-service agent said, 'Just a minute,' the astonished customer said, 'Boy, you guys *are* fast!'" Or groaned at stories like: "Two WestJetters were talking about the wonderful experience of flying. 'It's amazing how fast you can fly around the world now. My brother-in-law got to Singapore in less than twelve hours.' 'That *is* amazing,' said the second man. 'And what's even more amazing,' said the first man, 'is that he was trying to get to Moncton.'" Some flights have contests for little prizes: people dig into their pockets or purses to find the oldest penny on the airplane or try to guess the total of the flight attendants' ages. To win a game, some frisky passengers will even display a sock with a hole in it or bowl toilet-paper rolls down the cabin aisle.

A group called the WestJesters began at the airline's Regina station when customer-service agents devised a "tickle trunk" package of ideas to spread fun. It now has representatives at every airport who schedule bad-hair, disco, and other dress-up days for CSAs. "At Halloween one year," Calgary CSA Lynette Bryant says, "Glenn McPherson, then a CSA, now daily duty coordinator, dressed up as Princess Leia of *Star Wars* with two buns of hair and a gown and posed across our sales desk, completely still like a mannequin. The guests cracked right up, laughing."

Every airplane has a Just Plane Fun book. The WestJesters' inflight humour group compiles a thick, constantly updated binder of gags and games to help attendants amuse passengers. The manual warns, however: "Do not force humour. Realize that, for whatever reason, people may not be in the mood." On his flights with the airline, columnist William Thomas of the Welland, Ontario, *Tribune*, certainly hasn't been: "WestJet attendants are about as funny as a hard landing in Gimli, Manitoba ... As much as I admire the intent, WestJet is like a comedy club operating under the constraints of a No Laughter law." Sometimes the attempts at being funny can backfire badly. A male attendant trying to jolly a sombre-looking woman on one flight said flippantly, "I know you—you're that troublemaker." She turned out to be Dorothy Joudrie, a Calgary socialite who'd been acquitted of shooting her estranged husband. When she complained about the remark, the attendant wrote a letter of apology.

Don Bell says, "CanJet copied our business plan, right down to the most minor detail. But they said, in their infinite wisdom, people in Eastern Canada don't want humour on a flight, they want professionalism. So they took humour out of the mix." Even people without a sense of humour—but with the right name—might appreciate the occasional waggish promotions that Sandy Ruel's specialty-products group has overseen. They've included free flights for people named Black and Orange on Halloween, Heart and Love on Valentine's Day, and anyone with a Canadian PM's first and last name on the proposed Prime Minister's Day in February (which attracted ninety takers along with complaints of sexism because Canada has had only one female PM).

On board, even paying passengers tend to overlook the expected lack of free meals and relative shortage of legroom on the older aircraft when confronted by the easygoing and frequently entertaining ambience. You never know when a flight attendant might erupt in silly song like Tre Arnold, or

231

pull out a flute and play a beautiful melody like Shere Lowe, or more often, say something unexpectedly outrageous or funny, as so many attendants do (a Québécois male attendant, trying to ask passengers to applaud the last person to board, said, "Let's give him a round of the clap").

Once in a while, the airline has programmed professional entertainment on certain flights. A stand-up comedian who appeared one April Fool's Day proved a little raw for his captive audience, while a sleight-of-hand magician who turned passengers' $5 bills into $10 ones was a hit. Even more imaginative was WestJet's co-sponsoring of a published poet to write verse on board as part of a literacy campaign. Wendy Morton of Sooke, British Columbia, flew across the country in 2002, composing on the spot after chatting with passengers. One poem began with her admiring Jim, "a hunk who plays lacrosse," and ended with her knowledge that he had cancer—"Suddenly, I'm weeping/ because he knows more/about pain than I've ever known,/and joy."

But the airline's bedrock reputation is based on the quality of the reservation, check-in, and on-board transactions that customers have. The WestJet experience begins with booking a flight. Nearly 400 real-live reservations agents— oh, all right, Sales Super Agents—answer 85 per cent of customers' calls within a minute. If there's any delay, the call-waiting messages range from the informative (new flights, an eat-before-flying advisory) to the mildly interesting ("Did you know Newfoundland is the largest consumer of soft drinks?"). "We do not do hard-sell," says Donna Laitre, the sales agents' training adviser. "But it's imperative we give people all the options." The training course tries to hone soft-sell skills, which means "having a confident knowledge of products and making recommendations and creating a sense of urgency. If people hang up without buying this great product, you're doing them a disservice."

Increasingly, customers are avoiding any kind of telephone sell and booking by themselves or through travel agents via the Internet. "On New Year's Eve, 1999, somewhere around 3 per cent of our business was done on the Internet," Bill Lamberton says. Four years later, at least 70 per cent of all WestJet's reservations were being made on its internally designed Web site. In many ways, it's a model site. A user has an immediate incentive to reserve on line: a $6 discount per return trip and one Air Miles reward mile per $20 spent, both of which the company justifies by the significant savings it enjoys with Web bookings. Visitors can sign up for JetMail, which e-mails the latest seat-sale information, and seamlessly reserve hotel rooms and vehicle rentals with WestJet partners. A list of standard travel tips also includes advice on limbering up during a flight. A page that tells how to apply for employment, and get noticed doing it, also answers the question "why WestJet?" ("... one of the most unique corporate cultures in Canada," with profit-sharing, matching share purchase, cheap air travel, and other benefits). Sprinkled throughout the site are fun facts ("Did you know that Canadians invented such favorites as Yahtzee, Trivial Pursuit, Balderdash and Pictionary? For something to do on those long winter nights!") and sheer silliness ("How do trees get on the Internet? They log on").

Along with predictable sections such as corporate history and investor relations, the Web site has three unusual features: a monthly report of on-time performance; the company's mission, vision, and values statements; and an Interactive Feedback Corner. Customers can use the easy-to-navigate feedback option to rate and comment on all aspects of the airline's operation, especially with "suggestions on

233

what we can do to improve." The data are delivered to service specialists in real time, using customer-experience management technology developed recently by Vancouver-based ResponseTek Networks. ResponseTek immediately tabulates and grades every comment before sending it to the airline's client-support desk staffed by the call centre's four most experienced agents. They can resolve any problems themselves or pass them on to other departments or senior executives. Ro Imbrogno says that the online system has become a useful tool since mid-2002, informing decision-makers and front-line employees on a continual basis and letting them document the company's responses to passengers' reactions. Business manager Natalie Kerekes, one of those who studies the feedback, says it identifies areas that demand immediate improvement "and we can see trends where more investment is required to meet our guests' needs."

CUSTOMERS: HANDLE WITH CARE

Every year, WestJet receives thousands of voluntary expressions of written gratitude for the way employees have treated the people they call their guests. And sometimes the media pick up on a story of special human interest involving the airline. Here are a newspaper article and a passenger's note, from scores sampled, that give a flavour of the typical tone and content of the accolades:

Last Tuesday, Rebeka George of Williams Lake, B.C., was flying on WestJet with her three-year-old and her baby, from Regina to Prince George via Edmonton. As the plane descended into Edmonton she was racked with head pain, later diagnosed as a sinus infection. On landing, medics had her taken from the plane by ambulance to the Leduc hospital. Two WestJet ground attendants accompanied her and her children to the hospital, staying with her until she was discharged a few hours later.

Did they then go home? No, the attendants took the family to the Leduc Inn. Did they finally go home? No. Even though it was mid-evening, the WestJet attendants also checked in, taking an adjoining room to care for the family—including dinner for the family and going to the drugstore to fill a prescription for Rebeka. One attendant went home ... once another WestJet lady arrived to take her place. At the same time, WestJet located a friend of Rebeka's who'd been waiting in Prince George to drive her and the kids to Williams Lake. The friend was flown to Edmonton that same night, and also put up in the hotel.

Once the friend arrived, did the WestJet attendants call it a night? No, they stayed overnight in the next room, ensuring the George family and friend had breakfast, got to the airport, and were safely on the next WestJet connection to Prince George.

PS: This was all on WestJet's coin. "We were just amazed at WestJet's caring," says Rebeka's aunt Koren LeBlanc of Edmonton. "I got to the Leduc Inn by 8 p.m. that night. Even though I was there, the WestJet people still stayed. They were better than nurses!"

—*The Edmonton Sun*, September 29, 2002

Dear Mr. Donald Bell,

This letter is to inform you of my immense satisfaction with your great airline. On Wednesday October 3 [2001] I was scheduled to fly from Ottawa to Hamilton on your 6:40 pm flight. After checking in through security, which has become a bit of an angst-ridden experience, I was greeted by your most pleasant and courteous staff at the gate. After waiting a reasonable amount of time I was boarded for my—once again, on time—WestJet flight to Hamilton. After the flight attendants assured us that we were all nicely tucked in and the proper procedures for take-off had been followed, we pushed back from the gate. At that time it was discovered that there might be some engine trouble and all the passengers on board needed to return to the terminal building to wait. After a very short period of time we were instructed to move down to the WestJet counter to receive our complimentary hotel vouchers and transportation pass to the hotel, as we could not depart on that evening. I cannot tell you how helpful your ground people were. Inundated with all those people from the flight, they had to ensure that everyone was happy and comfortably set up for the night. A task that is not always easy with the tired traveler at the end of the day.

Now praise for the staff. They went out of their way. They remained polite and respectful to everyone, although that politeness and respect was not always returned. They worked quickly and efficiently to allow for everyone to move through the process as fast as possible. I cannot say enough about them and I am happy to tell you that I will certainly fly WestJet over and over. So kudos to your staff and thanks for being such a great airline ...

Yours truly,

Jon Martin[1]

For the past two years, laudatory comments from customers ballooned because of a clever seasonal campaign called Summer Smiles, which offered a chance to win travel gift certificates, a trip for two, and a digital camera. Passengers and WestJetters were encouraged to commend employees in writing for "wowing, surprising, or amazing" them. Customer-service director Ro Imbrogno brainstormed the idea with the marketing department, building on a routine public perception that "your people look like they enjoy their jobs." WestJet executives didn't need any elaborate PowerPoint presentation to convince them about the concept; she says they simply wondered, "Do you need more money to carry it on?" In summer 2003, the campaign spurred 2,571 letters that recognized 4,665 employees for exemplary customer service.

Does everyone deplane with the same warm feeling towards WestJet? Of course not—schedule delays or the odd less-than-pleasant flight crew can sour any passenger. And on rare occasions, employees or company policies can screw up badly. The handling of Margaret Mitchell is an instructive case in point. Because she is paralyzed down one side, her husband had called the airline the day before a Hamilton-Calgary flight in 2000 to pre-book two of the roomier bulkhead seats for him and his wife. After being told that they could arrange this at the airport, the couple found the requested seats were already assigned and they got standard seating. A flight attendant on board refused to exchange their seats; a deadheading pilot and his father were in the bulkheads. The Canadian Transportation Agency, hearing the complaint, ordered WestJet to amend its seating policy to ensure that the needs of disabled people are discussed during booking and check-in; that such passengers are pre-assigned the specific seats they request; and that those with disabilities get priority over able-bodied passengers. The case should be taught in all of the company's training sessions as an example of what not to do.

Yet complaints are surprisingly few, in large part because of WestJet's Customer Care department and a group of at least 600 employees across Canada who have volunteered to be Caregivers. In March 2003, a woman from Abbotsford, British Columbia, was flying back on the airline from Saskatoon when she fell ill as the airplane approached Calgary. A WestJet employee called the passenger's husband at home and asked if he could fly immediately to Calgary— "We have a ticket arranged for you." On arrival, he was met by a customer-service-agent Caregiver and a team leader, who drove him to his wife's hospital and stayed with them until after midnight, when her condition had improved. They then took the couple to a hotel at the company's expense and gave them meal vouchers and confirmation numbers for their free flight home the next day. The agent also left them his phone number with instructions to call during the night if they had any problems or in the morning if they needed help with their luggage. The wife was fit to travel and they arrived home safely. A month later, a WestJetter called from Calgary to see how she was doing. "It's no wonder you guys are doing so well," the husband said, "with that kind of care and concern shown my wife and me."

That's what Caregivers do. A team of three led by Leslie Roll in the head office supervises the national network of employees who've agreed to go beyond their normal duties and handle emergency situations like this. While most of their work deals with individuals, they are trained to provide care in the most calamitous of circumstances, such as an air crash. A manual defines their role in a disaster as a liaison between the airline and survivors and family members, but cautions them to be a channel of communication, not a therapist, and an aide to help families take action, rather than acting for them ("Having greater control facilitates the healing process").

At this writing, there have been no major calamities involving WestJet aircraft. Passengers who encounter lesser problems meet members of the Customer Care department, managed by Dean Puffer, a brawny, broad-faced guy with a sociology degree from the University of Calgary. He came to the airline in early 1997, starting as a reservations agent and then working with his department's forerunner, the Help Desk. Today, Puffer seems committed to operating what he calls "a nerve centre—because you touch so many different departments of the company." Visiting the office briefly while on paternal leave to look after his seriously ill infant daughter, he talks about the work done by forty-three full-timers in the department and twenty-five sales agents who are cross-trained to do basic customer care.

The airline has guidelines for the most common problems. "If we've lost a bag," he says, "we pay them for it and go one-up and compensate them for their inconvenience—likely with a travel credit for a year and a $25 travel bag. There's also $25 a day for the first three days for incidentals. If the bag isn't found, the tariff is $250 and we give them a travel credit on top of that. The file gets transferred to us and we start talking about the value of what was in it."

If a flight is delayed, "the WestJet promise is that if a plane breaks down and you have arrived two hours late, we'll give you a base-fare compensation—for a $99 ticket, a $99 travel credit. If it's weather, we can't. But it doesn't mean we won't take care of you with hotels and ground transportation. You don't have to ask for it or yell at us. And you'll likely get a letter from us within a week."

239

As many as sixty e-mails and fifteen letters flow into Customer Care each day, about half of them problems that need handling (the e-mails are answered within two hours, a time gap Puffer wants to shorten). Very occasionally, customers initially suggest that they'll sue the airline, go to the media, or—in one case—threaten to kick Puffer in the unmentionables. The manager remains even-tempered about irate or sometimes-unreasonable claimants and cautions his staff: "When people come to you asking for something, it may be natural to think they want to screw you—but you *can't* think that way. Probably less than 1 per cent of them aren't honest. Don't deal with the exceptions; deal with the general, who are not out to get you. And if you happen to find the person did lie or cheat, oh well, that's the cost of doing business."

Taking care of special passengers, he says, can be the biggest challenge, but also most rewarding. "A senior mother was travelling from Regina to Calgary and there was some confusion about what was needed for assistance; the senior wasn't guided as they expected. She was very scared and overwhelmed in the Calgary airport. Karli Davidson, a customer-service agent, came in on her day off and met her with some flowers before her flight home. And when she got off the plane in Regina, she was well attended to. The senior's daughter wrote saying how special she felt."

WestJet also has an unadvertised program for potential customers who have a fear of flying. It's run by Herb Spear, the original flight-operations supervisor, who is still spryly on staff at the age of eighty—probably the oldest airline employee in North America. Spear also helps run the occupational health and safety section and gives guided hangar tours, a range of tasks that explains why he starts at 5 a.m. and works twelve hours a day. (In 2003, he took one of his rare holidays to complete a five-kilometre run at the World Airline Road Race in Florida.) The reservations department and local hospital and university psychology departments refer fearful flyers. He tours them through a jet, explaining its operations and the source of the funny noises they might hear, and then takes them on a flight to Edmonton and back. "The thing that blows their minds is that we don't charge for this," Spear says. Of the forty or so people he's helped, only two remained fearful despite his efforts.

CARE AND FUELLING OF EMPLOYEES

Dean Puffer echoes so many of his colleagues in summing up what WestJet defines as its central philosophy: "Just treat our people well and our people will treat our customers well" (although, knowing how a handful of customers can behave, he adds, *"within reason"*). In the special corporate family that the airline tries to be, both the fun and the caring for colleagues start at home.

Within the workplace, WestJet uses CARE—Create A Remarkable Experience—as a special team under Don Bell's aegis to spread the culture and recognize employees for their achievements. A key component is the Fireside Chats that Bell initiated and other executives have adapted. He says, "You've got to think in terms of being a flight attendant and pilot who come to work and go into a crew room through a locked door, go out into a bus, and get into an airplane, and fly around Canada for four days. When do they get to touch the organization? They never connect. They have a lot of time on long flights to talk—it might be rumours, it might be bitching—and without any offset to that from management to say here's how we think, here's the Grand Why, our philosophy, where we're going, here's what's going on."

Each group of employees, from reservations agents to flight crews, attends at least one chat a year. They are paid

241

for their time and encouraged to ask questions of executives who share their visions of the company and the industry. And for employees to get recognition of their contributions. "Whether you're a pilot or a mechanic or in the call centre, after you've taken that 914,000th call, you can get burned out. It's human nature. Therefore, things like Fireside Chats [are vital] because you reconnect, you say, 'What I'm doing here is important because I've been here five years and I have a lot to give to the organization.' Celebrations and pats on the back are so important to those people."

Helping orchestrate things cultural is the ebullient Kristen To, the creatively titled "CARE Initiative Creative Lead." The slight, delicately featured young woman, who came to Canada as a four-year-old Vietnamese immigrant, studied psychology at the University of Calgary, with an emphasis on organizational behaviour. Explaining why the Big Shots decided to promote CARE in the company, she says, "It's the branding of our culture. They realized that our culture is just as important as marketing and other departments. Clive and Don always say executive actions will define where our culture will go, but it's the grassroots that keeps it moving."

She's involved in the Summer Smiles campaign and Culture Presentations, and meets monthly with a core group of eight managers from different departments to discuss employee-recognition tools and tactics. In one initiative, CARE mounts a changing show of cultural artifacts in the halls of head office: letters of praise and funny staff photographs, framed newspaper articles and some of the airline's more aberrant ads (one showing a hockey player's face flattened against the glass below the headline "Now boarding"). Various departments have their own CARE community. For instance, "the pilots have built a very sophisticated mentoring

242

program to celebrate all the graduates, take out-of-town pilots around Calgary, and describe how important our culture is ... In the call centre, Bernie and Muriel Monk, husband and wife, celebrate the birthdays of every one of the department's 400-plus people with a card and a little gift—out of their own pocket. They have not missed one person."

From the beginning, WestJet's people have marked their collective milestones ceremoniously. "When we made $300,000 a day," Ro Imbrogno remembers, "Clive bought us this big bell and every time we reached a milestone—even beat the week previous—we'd ring the bell. We learned to stop and celebrate our successes." Much later, the first million-dollar sales day prompted cake and a split of champagne, with a card signed by Beddoe, for everyone in the call centre. The company's own birthdays and twice-yearly profit-sharing distribution have become good excuses for big parties. At the anniversary celebrations, outstanding WestJetters receive Best of the West awards for being the funniest, friendliest, most unpredictable, or most thoughtful, and for going out of their way to make a difference. Pilot Rod Nugent earned an Above and Beyond award after two unaccompanied minors missed their connecting flight during an IROP. As his citation explained, "Rod stayed with the kids and when it was decided to move the children on another carrier, he even tried to escort them on the flight to the final destination. Naturally, the other carrier would not allow him to accompany them, so he actually purchased a ticket for himself in order to get the children home."

To recognize the efforts of maintenance staff on overnight shifts, employees from all over the company have organized Dirty Bird airplane washes. It's a family affair with kids, and even the Big Shots are there wielding hoses and brushes to scrub the 12.5-metre-high 737s. A recent Summer

Smiles barbecue, also at the hangar, featured one measure-ment-solutions employee in a blonde wig as the victim in a dunk tank. Individual departments have their own events through the year, such as BeanLand's pumpkin-decorating contest and Nerd Day ("basically coming as we usually are," Sandy Campbell says). "Let's face it," accounting director Janice Paget says, "you don't hear about any finance depart-ment being the strongest purveyors of culture. But ours is like that. We truly do enjoy each other's company and usually do a quarterly goofy thing"—such as playing tag in a maze with laser guns. They have curling and softball tournaments and are recognized as the sweetest Christmas-carol singers in head office, spreading their music to other departments and even to passengers at Calgary airport. Sometimes the fun is spontaneous: "One Friday afternoon, our payroll manager said to some hard-working people: 'Put the pens down—we're all going to the movies.'"

Every quarter or so, departments in the home office will decorate their digs with bizarre themes, which is why a vis-itor might find the People Department looking like a Polynesian beach, with a sign saying "Welcome to the Tikki Lounge" and a suggestion of a grass hut, a paper palm tree, lawn chairs, and a table with sunglasses, a squirt gun, and a beach ball. Sometimes the decor is a surprise, as it was when quality-assurance director Russ White returned from Christmas holiday to find his office turned into a literal pig pen. Steve Ogle, chief maintenance officer, was evening the score for a prank by setting up a pen and borrowing a piglet from a local Hutterite colony.

Pranksters seem to abound at WestJet. IRIS is a long-limbed fabric baboon named by Alanna Deis for the People Department's information system. Twice the creature has gone missing, once with a ransom demand for "good donuts" for every department and the second with photos of IRIS in tourist spots around the world over two months. Deis was impelled to concoct Missing Baboon messages on milk cartons spotted around the headquarters. As payback, she and VP Fred Ring broke into the information-technology office on a Saturday and stole the suspected perpetrators' phones, keyboards, and mice. On Monday morning, the IT people were forced to write a grovelling letter of apology. Marketing's Bill Lamberton received his own ransom note after his precious green diary vanished and a note ordered him to be more lenient with time off and schedule more celebrations in the marketing office. "I said we're not giving into terrorism and the next day, a canister showed up with a little burnt fringe of one page of the book. But I held my own. The next day we were in an executive meeting and from the roof of the building, all of a sudden this book came down on a hook outside the window. Another time the terrorists were attempting to light a match to my book and actually set the fire alarms off in the building and a fire truck showed up. But finally, one day I just started playing hardball and privileges were removed and the book showed up in the parking lot."

The social clubs in various WestJet cities are a somewhat more formal expression of this collective corporate spirit. In smaller bases, they build things around potluck meals and other family events. Until 2004, the president of the Calgary 737 Fun Club was the blonde and smiling Lin Walker, a self-confessed workaholic who started part-time in the call centre before becoming the executive assistant to Clive Beddoe, Mark Hill, Scott Butler, legal specialist Shawn Christiansen—"and anyone else who needs assistance. Because I do so many of the social and cultural things, I have another team member to make sure my regular work gets done." The clubs, independent of the company, are

245

funded by employee dues and raffle proceeds, which even support the annual Christmas party. "If the company isn't spending $100,000 on it, we all share in the profits," Walker says like a true WestJetter. "And not everyone belongs to the social club, so it doesn't seem right that the company pay for all these events that don't give back to everybody."

Chronicling all this sociability and employee care—and the relentless progress of the company—is the lively-looking monthly newsletter *Jet Lines*, printed on glossy stock, but edited with folksy touches. Until recently the editor was the equally lively Sarah Deveau, who's written a well-reviewed book inspired by her own college days *(Sink or Swim: Get Your Degree without Drowning in Debt)* and who writes book reviews too, for the *Calgary Herald* and WestJet's contracted-out, monthly inflight magazine, *AirLines Magazine*. The twelve- to sixteen-page newsletter, Deveau says, "is for the people, by the people. Anyone can submit an article; we'll help them polish it. We want to have people share their stories and publish them in a fun kind of way." Although the occasional item by managers can be pretty turgid stuff, much of the material is couched in bouncy prose under the bylines of front-line people. Surveys and contests keep the publication interactive (a recent competition asked readers to write a story creatively explaining what they'd do with travel-pass prizes). Perhaps its most important pages are those devoted to the highly personal roundups of goings-on at WestJet's cross-country stations. Even if the people at one base don't bother to read the news from other locations, they should get the sense that they're part of a collective culture sharing common values. At least that's the idea.

EXPORTING AND EMBRACING CULTURES

The biggest challenge confronting WestJet is to keep its culture alive over the next few years as it more than doubles in size to become a $2-billion-a-year business. How Canada's dominant low-cost, low-fare airline does that will determine whether its costs continue to decrease in any dogfight with smaller carriers (CanJet, Jetsgo) and the cumbersome but still bothersome Air Canada. This becomes even more crucial with WestJet's professed new desire not merely to stimulate markets, but to differentiate its product enough to lure passengers from competitors with attractive extras—frills. Some employees express their fear that, by growing too fast, the company may not be able to maintain its culture. Typical is Dale Mitchell, the thoughtful terminal support manager who worked nearly three decades for Air Canada and worries what will happen if the Red Baron decides to shrink its domestic traffic. "The public would expect WestJet to fill the void. Will we have expanded too fast too quickly and will we lose the essence of the WestJet culture—the western hospitality and ability to wow and amaze the guests?"

Since 2000, the westerners have had instructive experience in exporting their own corporate culture to far-flung outposts while dealing for the first time with a very different culture—that of French-speaking Canada. They began hiring bilingual flight attendants and reservations and customer-service agents and by 2003 had their first in-house communications and translation coordinator, Monica Schael. A Canadian air-force brat, with French as her first language, she lived across Canada and overseas before getting a journalism degree from University of King's College in Halifax. Along with translating WestJet documents, from news releases to Web-site pages, she works with qualified translators and in a pinch can be a spokesperson with Québécois media and coach company executives who have limited French. Her biggest challenge has been maintaining WestJet's friendly communications style: "French is much

247

more of a romantic and formal language, and this can pose great difficulties for me when translating documents."

Lamberton, when he was marketing and sales vice-president, could spend ten minutes explaining, and justifying, the WestJet approach to bilingualism. Boiled down, it was: "A price tag of $59 speaks both languages. The dollar sign is the same whether it's French or English. But on the customer-service side, we make much effort for any airplane that goes through Montreal to have at least one French-speaking flight attendant on board ... The fact is, we have our culture too and it's the western Canada culture, and if we can go into Montreal and say, 'Forgive our French—we're going to give you good service and a good price, and we're going to try hard.' ... But there certainly is a need for us to be aware that this culture needs special attention and make sure we do ask for forgiveness and try our best."

The proof was in the doing: opening WestJet's first east-coast base in the medium-sized New Brunswick city of Moncton, more than a third of whose population is French-speaking. Two company recruiters went there in early 2000 and found Lynette Lambert, an airline veteran of fifteen years who'd been station manager for Inter-Canadien, a feeder carrier for the vanished Canadian Airlines. The thirty-five-year-old Lambert, whose first language is French but whose English is flawless, was eager to do the same job for WestJet. She spent just a day and a half in Calgary meeting airports people, including director Dale Tinevez, who advised her: "Always try to pass on the glory to your people—let them have ownership." Then she and seven bilingual customer-service agents had two weeks' training in Moncton with the spirited Calgary reservations manager Pino Mancuso, who "lived and breathed all the WestJet values." They finished on a Friday and opened the station on

Monday for six flights a week (a frequency that has risen to twenty-nine in summer). Lori-Rae Bennett, a team leader from Kelowna, who knew the WestJet culture, moved to Moncton at her own suggestion—"as a way of ensuring the culture lived here," Lynette Lambert says.

Early on, the company suggested that the CSAs, working part-time, could increase their hours if the airline opened a satellite call centre in Moncton specializing in French-speaking booking agents. Although the experiment survived a year, it was found that having the same capacity in head office was cheaper. Lambert agreed, saying, "We wouldn't want to gain benefits locally at the expense of WestJet globally. It's one for all. The bottom line is very important to us." It's obvious that she has attended one of Bell's Culture Presentations in Calgary, as have all the newest Moncton CSAs, who now spend three weeks training there.

Has the WestJet spirit survived in translation—in the eastern reaches of Canada and in French? Lambert lists some of the Moncton people's random acts of fun and kindness: CSAs offering small prizes to customers during check-in for answering questions such as "How much do the two of us weigh?"; singing boarding announcements to the tune of "Limbo Rock"; and welcoming flight crews with little treats "just because it's a nice thing to do." How quickly many of the New Brunswickers have captured the spirit is illumined by some ... well, truly *touching* stories from the Moncton station. They arranged to have a penniless young couple fly free to Niagara Falls on their honeymoon. When a middle-aged woman was on a delayed flight and missed her bus connection to Prince Edward Island, a nineteen-year-old male CSA volunteered, without being asked, to drive her across the Confederation Bridge to PEI, a trip that takes a couple of hours each way.

But the most poignant tale involves a westerner who'd left his wife in anger and, thinking he was flying to St. John's, Newfoundland, landed in Moncton instead on a WestJet flight connecting to Saint John, New Brunswick. Michele (Mitchie) Bastarache, a compassionate customer-service agent, sat down with the distraught man and said, "I don't know what you're going through, but ..." Letting him talk, being sympathetic about his plight, she calmed him down so much that he decided to head back home. The Moncton staff arranged to fly him back to Calgary and gave him hotel and meal vouchers, all at WestJet's expense. A month later, they received a letter from his wife, saying she wanted to thank them very much, and while she didn't know what they'd said or done, it had made a difference.

Lynette Lambert says, "The one thing other airlines might be able to change is the one thing that's absolutely free: treating their customers with care and giving them good service. It seems so fundamental, yet is missed so often."

As Don Bell says, "If you took away the hospitality, the customer service, the people component of WestJet, what would we have? We'd have a bunch of airplanes. We would have Air Canada." But has transplanting such corporate culture in eastern soil proved at all difficult? "That has probably been the easiest thing we've done. Our philosophy is these new bases are microcosms of WestJet, mini-WestJets, their own entity—there's only five to twenty people and they create their own family. They hire the right people. We put a few people in there from other bases. [The new people] come out here for three weeks of training, and they get culture—culture is now embedded in the training. You're going to emerge from that really understanding what we value and what WestJet's all about, and go away feeling proud and connected. The airports have done an absolutely fabulous job. I think the culture in a town like Gander, which is about as far away from Calgary as you can get, is just as strong—if not stronger—than it is right here."

EPILOGUE
Flying into the future

"I could see WestJet becoming half the market for sure," said Cameron Doerksen, an analyst with Dlouhy Merchant Group Inc. in Montreal ... "WestJet has a good couple of years of growth ahead of them. Air Canada has to find a compelling reason to persuade people to return," said Douglas Reid, an airline expert at Queen's University in Kingston.
— *The Globe and Mail*, November 11, 2003

Air Canada chief Robert Milton said yesterday that the airline must continue to "aggressively" drive down its operating costs to compete with its discount rivals, as the airline reported a third-quarter loss of $263 million.
— *The Globe and Mail*, November 27, 2003

On the very day in August 2003 that fifty-five new WestJetters were making mock of Air Canada during a Culture Presentation at Calgary's Coast Plaza Hotel, they'd received one of Clive Beddoe's regular messages to the troops. While many of them hadn't yet met their CEO, they could infer much of his character and the corporate culture that he personified in these "Comments from Clive":

Yesterday Air Canada announced a $560 million loss for its second quarter. This is Air Canada's largest quarterly loss in recent history that contrasts sharply against our profit this quarter of $14.7 million, our second largest quarterly profit to date. Their results are obviously terrible and without a doubt demonstrate once again that their business model is completely broken.

With losses of this magnitude there is little doubt that their current restructuring efforts are going to involve even more layoffs and downsizing, which in turn will affect many of our peers in the industry and possibly some of our friends and family. Even more unfortunate is that much of Air Canada's losses were preventable, and fundamentally stem from the failed business model which their Board of Directors and Executives have clung to for years.

These losses, and the damaging effect they will have on Air Canada's employees, shareholders, and customers, underscore the incredible importance to WestJet that our people must continue to work together to maintain and build our successful team. As with any growing organization, we may not make the best decisions every time or be able to fix every problem we encounter, but the power of our great team has brought this company to where it is today and our team must continue to be the most important thing for us to focus on. You only have to look at the wealth that we have all created for our shareholders and for each other. In the last few weeks we have seen our share price increase from $16 to over $21 and this is a direct result of the strong financial earnings that we have all achieved.

Our climbing share price comes from the power of our ability to work together and we must do everything we can to ensure that we continue to sustain this power. We need to look at all that our team has accomplished and focus on the positive impacts that this unique working environment has brought to us all. Let's continue to keep communication channels open throughout all levels of our company and not let minor differences upset the overall positive environment that we have all worked so hard to create. Together we can find solutions to any challenge and it is the strength of our ability to find solutions that is the key to our ongoing success.

Air Canada's announcement only serves to emphasize that we have an enormous opportunity ahead of us. We need to remain focused on our successes and continue to work together to take advantage of the opportunities before us.

Beddoe would have another go at Air Canada two months later in a newspaper interview, claiming that the carrier was "pursuing something of a scorched-earth policy towards the industry ... It's obviously not working with us. We're continuing to prosper. But I suspect that it's doing considerable harm to their own earnings numbers and their own cash position ... It's absolutely ridiculous for Air Canada to think that they can sell seats at a third of their costs and make money doing it." Some outside observers, and at least one of his senior colleagues, thought he verged on sounding arrogant, and the *Globe and Mail*'s Alberta correspondent Deborah Yedlin came right out and wrote: "Mr. Beddoe couldn't help taking another shot at Air Canada ... [he] might crow all he likes ... a CEO of a company whining ... that's simply poor sportsmanship."

When I mentioned the column to him, Beddoe retorted, "She completely misunderstood my point: our society inherently rewards the weak and failed and penalizes the successful ... She took it as if I was crowing. That's not what I meant at all. But unfortunately if you're going to be outspoken, you're going to be misunderstood." Still, there was obvious danger in adopting too aggressive an anti–Air Canada stance: at what point does an underdog start looking like a bully and seem to be kicking a competitor down for the count? Far better, even his colleague believed, to stay publicly humble and focus on improving the product and lowering costs while keeping employees and customers happy.

To do that, it was vital to propagate the WestJet culture—which is what the CEO was now doing among admirers on this steamy Friday afternoon in Calgary. Beddoe is generally a contained individual, but in public appearances his big smile blossoms. An observant employee originally had reservations about him, thinking he must be brilliant because "he had created a culture of smiling, happy people when he wasn't one himself. He doesn't seem to enjoy the spotlight, though the media has it trained on him 24/7, and is very reserved in public, and even one-on-one. However, as a thank-you to a small team of people for their work on a project, he recently went to dinner with this group, along with Don Bell, for two hours. During the dinner, and in subsequent interactions, he was funny, outgoing, expressive with his hands and face, and extremely engaging. Once he's opened up with you, he's an extremely friendly guy. His body language opens, and the camaraderie is suddenly evident in every interaction."

Speaking to the new WestJetters, he had the same low-key charismatic quality about him. "What do you want to talk about?" he asked them, standing with a hand-held mike in the middle of the conference room. Air Canada's huge loss came up immediately. "I hate to say we're partly responsible for that. I want to reassure you we're just at the beginning of our development. Our biggest challenge is that they don't shrink too fast—ironically."

A man in the group began, "Are you still looking at ..."

"I'm going to stop you right there," Beddoe said. "It's *we.*"

Rephrasing, the questioner asked if "we" were still looking at expansion in Canada before going into the United States.

"The best thing we can do is build a network and frequency in Canada. It's still a year to eighteen months away. I've been saying that for two years now. It all depends on how far Air Canada shrinks."

Beddoe told them that WestJet, meanwhile, had options on fifty-four more Boeing aircraft. Among other things, the airline was increasing its charter business, which would rise to $36 million in 2003 from $16 million the year before. And it had just signed a $29-million deal to have Montreal-based Transat AT's tour companies, World of Vacations and Air Transat Holidays, use WestJet 737-700s and crews to fly as many as eighty thousand vacationers to the Caribbean in the coming winter. "The good thing about the charter business is it gives us incremental use of our aircraft ... and better utilization of our crews. It's very profitable for us to do the charter work. And it's a very good way to expose people to WestJet who wouldn't otherwise fly us."

Rather than answering a question about the new aircrafts' Required Navigation Performance capability, he turned it over to Bob Urban, a first officer who'd recently come from Air Canada's Jazz. "RNP will take all those nonprecision approaches across the country"—450 of them, Beddoe prompted him—"and turn them essentially into precision approaches. It's going to give us an operational advantage over a lot of other operators."

257

Then, ranging rapidly over several subjects, Beddoe mentioned the recent credit line of $100 million that Ontario Teachers' Pension Fund had extended the airline—"the luxury of it is that we don't *have* to use it." Asked his opinion of rival Jetsgo, he replied, "I'm surprised Jetsgo hasn't Jetsgone. Their seat cost is, by our estimation, higher than ours by about 20 per cent. I don't see how you can make it work." And, answering a query about the future of the Canadian airline industry, he said, "We're going to be the mass-market carrier and there'll be a high-end carrier ... I think we have the potential to get 60 per cent of the market."

With the questions dwindling, Beddoe thanked the new recruits and said as he left, "Have a very good weekend and a good summer."

FLYING INTO THE FUTURE

The summer of 2003 was a fine one for WestJet, as its net earnings for the third quarter—the twenty-seventh straight profitable quarter—increased about 40 per cent to $32.3 million from the year before. Meanwhile, its unit costs dropped to 10.6 cents from 12.6, largely because of its more fuel-efficient fleet of 737-700s. In the fall, the airline distributed more than $10 million to employees at a semi-annual profit-sharing party. After raising $150 million in an equity underwriting, in early 2004 it then arranged a $358 million (US) loan from the Export-Import Bank to buy more of the new aircraft and parts and install inflight satellite television. While intending to add eleven new Boeing 737-700s to its fleet throughout the year, the airline also began toying with the possibility of transcending its 737-only policy and buying smaller regional jets from the Brazilian manufacturer Embraer. This was in obvious reaction to the prospect of significantly increasing its domestic market share—perhaps to

Beddoe's hopeful 60 per cent—in the wake of Air Canada's battle out of bankruptcy protection.

The besieged senior airline announced in November 2003 that it had selected Trinity Time Investments, controlled by Victor Li, to invest $650 million as part of a $1.1-billion restructuring. Li, the eldest son of Hong Kong billionaire Li Ka-shing, has dual Chinese and Canadian citizenship and is the hands-on co-chair of Husky Energy in Calgary. Air Canada's board rejected a second approach by a rival bidder, New York-based Cerberus Capital Management. In January 2004 an Ontario Supreme Court judge approved the Trinity Time proposal. But as this was written, in early April, Trinity Time announced it was poised to walk away from its investment in Air Canada by month's end, having failed to gain concessions from the carrier's unions on the structure of their pensions. Although the investor's statement said "we do not rule out a continued participation if circumstances change sufficiently," even Robert Milton acknowledged he was not optimistic that Li would have a change of heart. As the Ontario Supreme Court extended Air Canada's protection from creditors to May 21, the company was seeking other investors, including Cerberus Capital, which wanted to discuss the new situation with the Airline's shareholders.

(The carrier then launched a $5-million lawsuit against WestJet, alleging that its rival had used the access code of a former Canadian Airlines employee, now a WestJet financial analyst, to illegally tap into an Air Canada website an astonishing 243,630 times and retrieve confidential data about profitable routes that would have given WestJet an unfair competitive advantage. The suit named the employee—who had left Canadian when Air Canada took it over but had access to a special reservation site— and co-founder Mark Hill,

VP of strategic planning, both of whom WestJet asked to take a leave of absence as it investigated the allegation.)

Jetsgo, CanJet, and Air Canada's Zip, with Steve Smith still at the controls, all had plans to expand. Jetsgo announced in early 2004 that it would more than double its fleet with eighteen Fokker-100s acquired from American Airlines. The 105-seat jets would supplement Jetsgo's fourteen 160-seat MD-83s. Its payroll of 600-plus employees now serving eleven Canadian and six American destinations would increase by 650. To spur demand, Jetsgo introduced a frequent-flier program and re-introduced its $1-a-seat Loonie Sunday promotion in the fall of 2003. Near the end of 2003, CanJet announced plans to triple its fleet by acquiring twenty new, longer-range airplanes over the next three years—and perhaps even to fly beyond its seven major centres in eastern and central Canada and three winter destinations in Florida. Earlier, it began dropping fares by as much as 40 per cent when WestJet began flying into the CanJet home base, Montreal, and Newfoundland. In a surprise move in November 2003, Zip pulled out of Montreal, Canada's second-largest market, leading the *National Post*'s Paul Vieira to report that "questions are raised about Zip's future, given that Air Canada eliminated its Tango-branded flights from its system." But in February 2004, Air Canada's Robert Milton said Zip would grow to twenty aircraft, with Airbus A319s replacing its older Boeing 737-200 fleet, beginning in June. And Smith said Zip might start flying into the United States. (Less of a threat to WestJet was a new Canadian carrier, HMY Airways, which marked its first money-losing year of regular scheduled flights to Toronto; Los Angeles; Las Vegas; and Mazatlàn, Mexico. Multi-millionaire Vancouver owner David Ho was intending to add four more aircraft to his fleet to fly trans-Pacific routes.)

Whatever happened, WestJet would be fighting for market share with its product-differentiation strategy and continuing emphasis on reduced costs through some technological advancements—but mostly through WestJetters' teamwork and productivity.

Meanwhile, the airline had to face four definitive decisions. One was whether to end its reliance on Hamilton as its eastern hub. In January 2004, the Canadian media were abuzz with the news that WestJet was switching 60 per cent of its Steel City flights to Toronto's expensive but more strategically located Pearson International. Central to the argument for the move were the convenience for business travellers and the fact that local business from Hamilton-area passengers had reached a saturation point.

"Sometimes you don't know you've overdone it until you *have* overdone it," Beddoe explained to me. "We added frequency in an attempt to create demand. The initial short-haul flights from Hamilton to Ottawa and Montreal didn't work. We added frequency to them to try to make them work and they did get a little better—but not much. So here we were, losing millions of dollars a year on these routes, not because they necessarily had poor load factors, but because—this is the strange part of this business—because the yield on each passenger was too low. And the reason the yield was too low was because the average number of local passengers we were carrying was very low. We were getting high through and connecting traffic, but we were getting very, very low local traffic. Through and connecting traffic has low average yield on it because the fare is spread out over the entire trip. Local traffic has much higher yield—but we were not able to attract sufficient local traffic on those routes." Pearson's high fees—which will cost the airline about $20 more per person—would be balanced by higher yields from the corporate flyers.

261

WestJet suffered a temporary setback in February when an Ontario judge ordered the Greater Toronto Airport Authority to give Air Canada preferential access to all fourteen gates at Pearson's newest terminal. The authority, which indicated it would appeal the decision, had wanted WestJet to share six of the gates with its rival. Beddoe said the expansion plan for Toronto was still on track and he hoped the company would eventually gain access to the more convenient gates as they become available.

Two weeks after announcing the increased flights through Toronto, WestJet revealed its decision to finally launch regularly scheduled service to the United States. Curiously, the news was buried in a media release announcing that it was adding 120 non-stop weekly flights to its Canadian schedule in the summer of 2004. Just a few months earlier, Beddoe had been saying that WestJet would be patient as it waited to see how the saga of the insolvent Air Canada played out. "If they shrink too fast, the void will be so big that we'll be really challenged [in serving the Canadian market]." Given that he now feels the major airline will remain heavily in the domestic market, the announcement of transborder flights was no great surprise.

Starting in October 2004, WestJet was to fly from various Canadian cities to Los Angeles and to Fort Lauderdale and Orlando, Florida, with seasonal service to Phoenix, Arizona, and Palm Springs, California. By taking a first small step and focusing on cruise-ship, theme-park, and golf and sun destinations, it wouldn't be competing directly with Air Canada's mainly business-traveller clientele flying between the two countries. In recent years WestJet has been operating charter flights south of the border for Transat AT of Montreal. Beddoe calls the airline's own scheduled service "a prudent, logical next step—we think there's a huge market there." (In March 2004, JetBlue Airways—founded by David Neeleman,

262

one of the original investors in WestJet—filed a federal application in the US for a certificate allowing it to fly into Canada and four other countries. Although there were no immediate plans for Canadian flights, Neeleman has said that Canada is part of the his airline's expansion plans. He no longer has any financial interest in WestJet.)

Just before WestJet reported its plan for the US market, the *Globe and Mail*'s Keith McArthur had wondered, in an article in *Report on Business Magazine*, whether the carrier was extending itself too far, too fast—overexpanding in a way that had crippled Wardair and Canada 3000. Beddoe told me he doesn't think so: "We are an engine of change; we bring about change to an industry as a low-cost provider that is essentially displacing high-cost capacity. And in order to do that, you have to push the envelope. So do we over-supply markets temporarily? Probably. But do we do it profitably? Well, apparently, we had one of the best margins in the industry last year. The luxury of the business is that you can redeploy capacities. And that's what we're doing." McArthur himself, recognizing that operating costs are the key element in surviving as a low-cost airline, noted that WestJet's unit costs had recently dropped by 16 per cent: "These numbers suggest that any concern about WestJet allowing its costs to creep up may be unfounded."

Another decision that faced Beddoe and his directors had been in gestation for a long time. To satisfy corporate-governance concerns that have been surfacing so strongly throughout North America, WestJet appointed another independent director to its board in late 2003. Allan Jackson of Calgary is president and CEO of Jackson Enterprises, a holding and consulting company, and of ARCI Ltd., a real-estate investment company (and the Canadian holding company of the d'Arenberg family, whose investments are mostly in

Europe and South America). He's been exposed to a wide range of businesses in his role as head of the investment committee of Canadian Western Bank, which has yielded another WestJet director, bank president Larry Pollock. Jackson represents what Beddoe describes as "people who have no vested interest, no stake [in the company] in the classic sense"—and who, unlike the other directors, would likely be compensated for their services.

Then just one day after the board appointment, the first long-term member of the founding group of executives took his leave from WestJet. Bill Lamberton—an industry veteran for a quarter of a century, who'd been a partner in a travel agency and worked for three other airlines—stepped down as vice-president of marketing and sales. It wasn't an easy decision for any of the parties involved. All but a handful of WestJetters were caught by surprise. His going was handled with some grace and much sensitivity—to him, his colleagues, and share-holders inside and outside the company. Clive Beddoe's public farewell read: "It is with regret that I announce that Bill Lamberton will be leaving WestJet. Bill has been with WestJet since our inception, and has contributed greatly to our success over the past eight years." It was suggested that the forty-nine-year-old Lamberton might play an ongoing role as a consultant. The surface amicability of the leave-taking was very WestJet-like, enough to reassure the first-responder industry analysts. Cameron Doerksen of Dlouhy Merchant Group said, "I don't think it will be a problem. WestJet has an experienced manage-ment team." Lamberton himself did a token interview with the *Calgary Herald*, in which he said that while the company was "exciting and vibrant," he'd decided to step back to explore other opportunities.

Shortly after, he told me, a little enigmatically, "It's eight years in a very pivotal position in a driven organization, a job I love, but there's a time when you say, 'Can I do a little less?' I've had a good run with a company that celebrates success and makes a difference for people. I worked for a lot of airlines in my time and it's been a very good stint for me ... I don't have any hard feelings about leaving the company. I'm okay financially and okay to step back. My associates at the executive table are supportive—and are going to ask me for advice. The consulting is very real." He talked of being the first senior alumnus of the founding executives who could bring a WestJet slant to industry boards such as Nav Canada, the civil air–navigation service. Summing up, he said, "There was an opportunity placed before me and I was happy to take that route."

If there was any disagreement between him and Clive Beddoe, did it have anything to do with the future direction of the rapidly expanding airline? All Beddoe would say is: "Bill did a great job for us." (And on learning that WestJet was to receive a Marketer of the Year Award in Vancouver at the Academy of Marketing Science's annual conference that May, he asked the former marketing and sales VP to accept it.) Whatever the underlying reasons for Lamberton's departure, his absence left a gaping void in the company, which would have to be filled as soon as possible.

Surprisingly, the quiet drama of his exit had occurred only days after he'd attended a crucial gathering of the four founders, People vice-president Fred Ring, and CFO Sandy Campbell. They met at General Electric's twenty-one-hectare management-training centre in the Crotonville neighbourhood of Ossining in upstate New York. Jack Welch, GE's legendary former CEO, had developed the facility as the world's first major corporate business school (the WestJetters would agree with Welch's dictum to "use the brains of your workers," but not necessarily his concept of "motivating people by alternately hugging and kicking them").

Don Bell had been pushing hard for such a conclave. "We've built a very good organization," he'd told me a few months earlier, "and now we have to have some succession planning, strategizing the future, how we develop the structure—really planting the seeds for our succession ... The people I have working for me are exceptional, they're doing a fabulous job, and those people are ready to move up to higher goals. We need to go away and talk about that and think about that. We're trying to get that on the table. If we don't do it, I'm quitting"—although neither he nor I took that last statement too seriously.

Finally, everyone's schedule was arranged for two days of intense discussion on neutral turf in another country. "We were *there*—which says something about our ability to work together," Bell says with satisfaction. The vital topic of contemplating succession was on the agenda at Crotonville. Not short-term planning: as Beddoe points out, "If I was knocked down by a bus, we have a successor named, and so do the others." His vision of the get-together meshed with his co-founder's: "... to help crystallize our thoughts on what would be the prerequisites to get us to the next major hurdle, to be a $2-billion business—the qualities and skill sets of people in the organization to help us get there ... And what sort of qualities would we need in people who would succeed us?"

The seven men were emboldened by a facilitator who told them that many companies never take the time to do such planning. While they talked about the next two critical years, Bell says, "We had a good chat about how we've all put our heart and soul in and we may never really go away. How would I be able to quit and have nothing to do with WestJet?" At one point, they broke into two groups to brainstorm about which values truly define the company.

266

Reconvening, they were pleased to learn they'd all used strikingly similar words.

What the Big Shots decided on as they approached their eighth anniversary on February 29, 2004, was, reassuringly, pure WestJet. In the end, they expressed the values they prize as the three E's—a succinct rephrasing and reaffirming of the company's original vision. It's a vision that will be tested persistently in the coming years as this sleek, airworthy Canadian company flies through the turbulent skies of the North American airline industry into an auspicious future:

Emotional commitment of WestJet's people

Empowerment

Entrepreneurial attitude

—all of which drive profit, productivity, safety, and great customer service.

A BRIEF HISTORY OF WESTJET AIRLINES

1994

With backing from Calgary businessman Clive Beddoe, employee Mark Hill researches models of Southwest Airlines and other no-frills carriers in the US, identifying effects of market stimulation on airline success. They and two other local entrepreneurs—Don Bell and Tim Morgan—become partners in a venture to launch WestJet Airlines, Canada's first low-cost carrier, which intends to empower employees with financial incentives.

1995

WestJet office opens quietly in Calgary after founders raise $8.45 million from eleven investors and begin assembling a team of employees, including pilots from other airlines. American David Neeleman, former executive of Southwest and discount airline Morris Air, also invests and offers advice along with Dan Hersh, a consultant to Morris Air and ValuJet. Ontario Teachers' Pension Plan invests $10 million for a 24.2 per cent stake. Bill Lamberton of Canadian Airlines joins to oversee marketing, sales, and scheduling.

1996

Announces its birth to the public on February 5. With a roster of 220 employees on February 29 (including controller Sandy Campbell from Canadian Regional Airlines), begins serving Vancouver, Kelowna, Calgary, Edmonton, and Winnipeg with three used Boeing 737-200s. Within six months, is flying to Victoria, Regina, and Saskatoon and has carried 388,000 passengers. Low-cost competitor Greyhound Air begins service in May (but will be gone by September 1997).

On September 17, Beddoe and colleagues shut down the airline voluntarily after Transport Canada refuses to let it continue operating under an aircraft maintenance program the government agency earlier approved. Employees—none of whom are laid off—work around the clock to sastify federal regulations; seventeen days later, relaunch to great support from the public. By year's end, flies a half-million passengers. Despite the shutdown, records ten-month revenue of $37.2 million and nominal profit—with each employee receiving $500 in profit-sharing.

1997

Carries one-millionth passenger and pioneers use of regional airport in Abbotsford, BC, for regularly scheduled service, bypassing more expensive and crowded Vancouver International Airport. Don Bell, backed by Beddoe and other partners, deepens corporate-culture offerings.

1998

Revenues rise by 63 per cent to $125.9 million, generating $6.5-million profit, as fleet of 737-200s increases to eleven from six. Employees average $2,800 in profit-sharing. Air Canada and Canadian Airlines International collectively lose more than $150 million over the year. Canadian Union of Public Employees fails in attempt to unionize WestJet (following unsuccessful attempt by Canadian Auto Workers in 1996).

Team including Tim Morgan "liberates" WestJet 707 from California aircraft overhauler and flies it back to Calgary in clandestine operation; later lawsuit against airline dies after US company goes bankrupt.

1999

Steve Smith, president of Air Canada affiliate Air Ontario, becomes WestJet president and CEO. Initial public offering in July totals $25 million at $10 a share; stock closes at $18.65 by year's end. Company matches contributions of WestJetters investing up to 20 per cent of their income in share-purchase plan. Employees create their own organization—PACT, the Pro-Active Communication Team—to bring their concerns to management. Passengers can now book flights on a WestJet Web site. After floundering, Canadian International merges with Air Canada.

In February, WestJet and Air Canada announce talks about possible profit-sharing venture to have WestJet handle short-haul flights; plan dies after Air Canada swallows floundering Canadian Airlines. In December, after five-millionth passenger recorded, WestJet announces plan to enter eastern Canadian market.

2000

Between March and early June, service begins to Hamilton (rather than Toronto's pricy, congested Pearson International), Moncton, and Ottawa. After Steve Smith is asked to leave, Clive Beddoe returns as CEO. Former high-school principal Fred Ring becomes vice-president, People. Annual earnings have increased by 143 per cent. Now flies twenty-two aircraft and plans to acquire thirty-six new-generation Boeing 737-700s over next five years, with loan guarantees from US government's Export-Import Bank.

Moves into new headquarters near the Calgary airport; a new hangar will open early in 2001 for up to six aircraft, flight simulators, and maintenance, operations, and inflight teams. Starts a CARE initiative—Create A Remarkable Experience—to strengthen corporate culture and recognize employee achievements. Profit-sharing pool reaches $13.5 million, giving average employee more than $9,000. The four founders are named Canada's Entrepreneurs of the Year.

2001

First four 737-700s arrive, with winglet technology to save significantly on fuel costs. Airline enjoys twenty-first consecutive profitable quarter despite travel disruptions after 9/11 terrorist attacks, when employees rallied to assist beleaguered US airlines in Calgary. Year's load factor (number of seats occupied) is a healthy 74.7 per cent. Fifty thousand people apply for jobs with the carrier; 500 are hired. Payroll grows to 2,300 and 82 per cent of employees are in voluntary share-purchase program. WestJet named one of the top 100 employers to work for in Canada and founders visit Monaco to receive International Teamwork Award.

Canada 3000, after acquiring CanJet Airlines and Royal Airlines, goes bankrupt. CanJet later resurfaces in Halifax and ex-CanJet head launches low-fare Jetsgo in Montreal. Air Canada starts a discount subsidiary airline, Tango.

2002
Service begins to Toronto and London, Ontario, as fleet grows to thirty-five, including 737-700s specially fitted with leather seats and larger stowage bins. Air Canada and many US airlines seek bankruptcy protection; WestJet's revenues increase by 42 per cent to $680 million, with a $51.7-million profit. Completes $82.5 million common-share offering. Workforce expands by a third to 3,120, including 120 in first company-run ground-handling crew, operating in Calgary, where WestJet moves into roomy new airport wing and introduces Canada's first bridge to allow simultaneous front and rear boarding.

Steve Smith becomes president of Zip, Air Canada's new Calgary-based, low-cost carrier. Air Canada Jazz launches as Halifax-based regional subsidiary with lower-paid pilots flying turboprop aircraft into small communities.

2003
Unit operating costs drop to 10.6 cents from 12.6. WestJet begins flying to Windsor, Ontario; Montreal; Halifax; and St. John's and Gander, Newfoundland. Charter business grows to $35 million from $16 million in 2002. Signs deal with Montreal-based Transat AT tour companies to use WestJet airplanes and crew on Caribbean flights. Announces plan to increase legroom on aircraft and in 2004 offers first real on-board frill—satellite TV on personal video screens—and will be the first Canadian carrier using required navigation performance system to pinpoint aircraft position independent of air-traffic control.

Air Canada reports $263-million third-quarter loss while seeking a major investor (who will be Victor Li of Trinity Time Investments). WestJet reports quarterly net-earnings increase of nearly 40 per cent to $32.3 million and distributes more than $10 million in semi-annual profit shares to employees, whose number now approaches 4,000.

2004

As Air Canada Jazz, Jetsgo, and Zip announce expansion plans, WestJet switches 60 per cent of its Hamilton traffic to Toronto's Pearson International and decides to add 120 non-stop weekly flights to its Canadian summer schedule. Two weeks later, announces long-planned move into US with selected flights to cruise-ship, theme-park, and golf and sun destinations. Clive Beddoe calls it "a prudent, logical next step."

NOTES

Chapter 2: Prepare for Takeoff

1. Freiberg and Freiberg, *Nuts!* (New York: Broadway Books, 1996).
2. Skene, *Turbulence: How Deregulation Destroyed Canada's Airlines* (Vancouver: Douglas & McIntyre, 1994).
3. Skene, *Turbulence.*

Chapter 3: Find Good Flight Instructors

1. Czarnecka, "Airline Cowboys: The Inside Story of WestJet," *Lexpert*, September 2001.
2. Newman, *Titans: How the Canadian Establishment Seized Power*, vol. 3 of *The Canadian Establishment* (Toronto: Viking/Penguin, 1998).

Chapter 7: Question Your Pilot

1. Glanz, *CARE Packages for the Workplace: Dozens of Little Things You Can Do to Regenerate Spirit at Work* (New York: McGraw-Hill, 1996)
2. Curran, The Canadian Press, May 6, 1999.
3. Collins, *Good to Great: Why Some Companies Make the Leap and Others Don't* (New York: HarperCollins, 2001).

Chapter 8: Fly in Close Formation

1. Gittell, *The Southwest Airlines Way* (New York: McGraw-Hill, 2003).

Chapter 11: Have a Fun Flight

1. From the personal collection of Don Bell.

INDEX

275